As the Cube Turns

Dan Fendel and Diane Resek
with
Lynne Alper and Sherry Fraser

Limited Reproduction Permission:
© 2000 by the
Interactive Mathematics Program
All rights reserved.

Individual teachers who have purchased this book for use with *Interactive Mathematics Program: Year 4* are hereby granted permission to reproduce student blackline master pages for use with their students. Reproduction for an entire school or school district or for commercial use is prohibited.

This material is based upon work supported by the National Science Foundation under award number ESI-9255262. Any opinions, findings, and conclusions or recommendations expressed in this publication are those of the authors and do not necessarily reflect the views of the National Science Foundation.

Unauthorized copying of the *Interactive Mathematics Program: Year 4* is a violation of federal law.

®Interactive Mathematics Program is a registered trademark of Key Curriculum Press. ™IMP and the IMP logo are trademarks of Key Curriculum Press.

Key Curriculum Press
1150 65th Street
Emeryville, California 94608
510-595-7000
editorial@keypress.com
http://www.keypress.com

10 9 8 7 6 5 4 3 2 1   03 02 01 00 99
ISBN 1-55953-346-3

Printed in the United States of America

Project Editor
Casey FitzSimons

Project Administrator
Jeff Gammon

Additional Editorial Development
Masha Albrecht, Mary Jo Cittadino

Art Developer
Ellen Silva

Production Editors
Caroline Ayres, Kristin Ferraioli

Project Assistants
Stefanie Liebman, Beck Finley

Copyeditor
Thomas L. Briggs

Cover and Interior Design
Terry Lockman
Lumina Designworks

Production and Manufacturing Manager
Diana Jean Parks

Production Coordinator
Laurel Roth Patton

Art Editor
Kelly Murphy

Photo Researcher
Laura Murray

Technical Graphics
Tom Webster, Lineworks, Inc.

Illustration
Alan Dubinsky, Tom Fowler, Nikki Middendorf, Evangelia Philippidis, Paul Rogers, Sara Swan, Martha Weston, April Goodman Willy, Amy Young

Publisher
Steven Rasmussen

Editorial Director
John Bergez

MATHEMATICS REVIEW
Lynn Arthur Steen, St. Olaf College, Northfield, Minnesota

MULTICULTURAL REVIEWS
Genevieve Lau, Ph.D., Skyline College, San Bruno, California
Luis Ortiz-Franco, Ph.D., Chapman University, Orange, California
Marilyn E. Strutchens, Ph.D., University of Maryland, College Park, Maryland

TEACHER REVIEWS
Kathy Anderson, Aptos, California
Dani H. Brutlag, Mill Valley, California
Robert E. Callis, Oxnard, California
Susan Schreibman Ford, Stockton, California
Mary L. Hogan, Arlington, Massachusetts
Jane M. Kostik, Minneapolis, Minnesota
Brian R. Lawler, Carson, California
Brent McClain, Portland, Oregon
Michelle Novotny, Aurora, Colorado
Barbara Schallau, San Jose, California
James Short, Oxnard, California
Kathleen H. Spivack, New Haven, Connecticut
Linda Steiner, Escondido, California
Marsha Vihon, Chicago, Illinois
Edward F. Wolff, Glenside, Pennsylvania

Acknowledgments

Many people have contributed to the development of the IMP™ curriculum, including the hundreds of teachers and many thousands of students who used preliminary versions of the materials. Of course, there is no way to thank all of them individually, but the IMP directors want to give some special acknowledgments.

We want to give extraordinary thanks to these people who played unique roles in the development of the curriculum.

- **Matt Bremer** did the initial revision of every unit after its pilot testing. Each unit of the curriculum also underwent extensive focus-group reexamination after being taught for several years, and Matt rewrote many units following the focus groups. He has read every word of everyone else's revisions as well and has contributed tremendous insight through his understanding of high school students and the high school classroom.

- **Mary Jo Cittadino** became a high school student once again during the piloting of the curriculum, attending class daily and doing all the class activities, homework, and POWs. Because of this experience, her contributions to focus groups had a unique perspective. This is a good place to thank her also for her contributions to IMP as Network Coordinator for California. In that capacity, she visited many IMP classrooms and answered thousands of questions from parents, teachers, and administrators.

- **Bill Finzer** was one of the original directors of IMP before going on to different pastures. Though he was not directly involved in the writing of Year 4, his ideas about curriculum are visible throughout the program.

- **Lori Green** took a leave from the classroom as a regular teacher after the 1989-1990 school year and became a traveling resource for IMP classroom teachers. In that role, she has seen more classes using the curriculum than we can count. She has compiled many of the insights from her classroom observations in the *Teaching Handbook for the Interactive Mathematics Program®*.

- **Celia Stevenson** developed the charming and witty graphics that graced the prepublication versions of all the IMP units.

Several people played particular roles in the development of this unit, *As the Cube Turns*.

- Dean Ballard, Matt Bremer, Donna Gaarder, Steve Jenkins, Dan Johnson, Jean Klanica, Barbara Schallau, and Adrienne Yank helped us create the version of *As the Cube Turns* that was pilot tested during 1993-1994. They not only taught the unit in their classrooms that year, but they also read and commented on early drafts, tested almost all the activities during workshops that preceded the teaching, and then came back after teaching the unit with insights that contributed to the initial revision.

- Barbara Schallau, Adrienne Yank, and Kathy Anderson joined Matt Bremer for the focus group on *As the Cube Turns* in January 1997. Their contributions built on several years of IMP teaching, including at least two years teaching this unit, and their work led to the development of the last field-test version of the unit.

- Dan Branham, Matt Bremer, Steve Hansen, and Mary Hunter, field tested the post-focus-group version of *As the Cube Turns* during 1997–1998. Dan and Matt met with us to share their experiences when the teaching of the unit was finished. Their feedback helped shape the final version that now appears.

In creating this program, we needed help in many areas other than writing curriculum and giving support to teachers.

The National Science Foundation (NSF) has been the primary sponsor of the Interactive Mathematics Program. We want to thank NSF for its ongoing support, and we especially want to extend our personal thanks to Dr. Margaret Cozzens, who was Director of NSF's Division of Elementary, Secondary, and Informal Education during IMP's development period, for her encouragement and her faith in our efforts.

We also want to acknowledge here the initial support for curriculum development from the California Postsecondary Education Commission and the San Francisco Foundation, and the major support for dissemination from the Noyce Foundation and the David and Lucile Packard Foundation.

Keeping all of our work going required the help of a first-rate office staff. This group of talented and hard-working individuals worked tirelessly on many tasks, such as sending out units, keeping the books balanced, helping us get our message out to the public, and handling communications with schools, teachers, and administrators. We greatly appreciate their dedication.

- Barbara Ford—Secretary
- Tony Gillies—Project Manager
- Marianne Smith—Communications Manager
- Linda Witnov—Outreach Coordinator

We want to thank Dr. Norman Webb of the Wisconsin Center for Education Research for his leadership in our evaluation program, and our Evaluation Advisory Board, whose expertise was so valuable in that aspect of our work.

- David Clarke, University of Melbourne
- Robert Davis, Rutgers University
- George Hein, Lesley College
- Mark St. John, Inverness Research Associates

IMP National Advisory Board

We have been further supported in this work by our National Advisory Board—a group of very busy people who found time in their schedules to give us more than a piece of their minds every year. We thank them for their ideas and their forthrightness.

David Blackwell
    Professor of Mathematics and
        Statistics
    University of California, Berkeley

Constance Clayton
    Professor of Pediatrics
    Chief, Division of Community
        Health Care
    Medical College of Pennsylvania

Tom Ferrio
    Manager, Professional Calculators
    Texas Instruments

Andrew M. Gleason
    Hollis Professor of Mathematics
        and Natural Philosophy
    Department of Mathematics
    Harvard University

Milton A. Gordon
    President and Professor of
        Mathematics
    California State University, Fullerton

Shirley Hill
    Curator's Professor of Education
        and Mathematics
    School of Education
    University of Missouri

Steven Leinwand
    Mathematics Consultant
    Connecticut Department of Education

Art McArdle
    Northern California Surveyors
        Apprentice Committee

Diane Ravitch (1994 only)
    Senior Research Scholar
    Brookings Institution

Roy Romer (1992-1994 only)
    Governor
    State of Colorado

Karen Sheingold
    Research Director
    Educational Testing Service

Theodore R. Sizer
    Chairman
    Coalition of Essential Schools

Gary D. Watts
    Educational Consultant

Finally, we want to thank Steve Rasmussen, President of Key Curriculum Press, John Bergez, Key's Executive Editor for the IMP curriculum, Casey FitzSimons, Project Editor, and the many others at Key whose work turned our ideas and words into published form.

Dan Fendel   Diane Resek   Lynne Alper   Sherry Fraser

As the Cube Turns

Foreword

"I hated math" is an often-heard phrase that reflects an unfortunate but almost socially acceptable adult prejudice. One hears it from TV announcers, politicians, and even football coaches. I'll bet they didn't start their education feeling that way, however. According to the Third International Mathematics and Science Study, American fourth graders score near the top of their international peers in science and math. Surely mathphobia hasn't broken out by that grade level. By twelfth grade, however, students in the United States score among the lowest of the 21 participating nations in both mathematics and science general knowledge. Even our advanced math students—the ones we like to think are the best in the world—score at the very bottom when compared to advanced students in other countries. What happened? Is there something different about our students? Not likely. Is there an opportunity for improvement in our curriculum? You bet.

Traditional mathematics teaching continues to cover more repetitive and less challenging material. For the majority of students, rote memorization, if not too difficult, is certainly an unenlightening chore. The learning that does result tends to be fragile. There is little time to gain deep knowledge before the next subject has to be covered. American eighth-grade textbooks cover five times as many subjects in much less depth than student materials found in Japan. Because there is no focus on helping students discover fundamental mathematical truths, traditional mathematics education in the United States fails to prepare students to apply knowledge to problems that are slightly different and to situations not seen before.

As an engineering director in the aerospace industry, I'm concerned about the shrinking supply of talented workers in jobs that require strong math and science skills. In an internationally competitive marketplace, we desperately need employees who have not only advanced academic skills, but also the capability to discover new, more cost-effective ways of doing business. They need to design with cost as an independent variable. They need to perform system trades that not only examine the traditional solutions, but explore new solutions through lateral, "out of the box" thinking. They need to work in teams to solve the most difficult problems and present their ideas effectively to others.

Programs like IMP foster these skills and fulfill our need as employers to work with educators to strengthen the curriculum, making it more substantive and challenging. I can attest to the value of IMP because, as the father of a student who has completed four years of the program, I've discovered that something *different* is going on here. My son is

Interactive Mathematics Program

given problems around a theme, each one a little harder than the one before. This is not much different from the way I was taught. What is different is that he is not given the basic math concept ahead of time, nor is he shown how to solve upcoming problems by following the rule. By attacking progressively harder problems in many different ways, he often learns the basic mathematical concepts through discovery. He is taught to think for himself. He says that the process "makes you feel like you are actually solving the problem, not just repeating what the teacher says."

This process of encouraging discovery lies at the heart of IMP. Discovery is not fragile learning; it is powerful learning. My son thinks it can be fun, even if he won't admit it to other students.

I have another window on IMP as well. As the husband of a teacher who helped to pioneer the use of IMP in her district, I've learned that teaching IMP is a lot more than letting the students do their own thing. Lessons are carefully chosen to facilitate the discovery process. Points are given for finding the correct answer, and points are given for carefully showing all work, which is as it should be. Because the curriculum encourages different ways of solving a problem, my wife spends more than the typical amount of time teachers spend in reading and understanding students' efforts. The extra time doesn't seem to burden her, however. I think she thinks it's fun. She even gets excited when she sees that the focus on communicating and presenting solutions is measurably improving her students' English skills.

Let me conclude with a word of encouragement to all of you who are using this book. I congratulate you for your hard work and high standards in getting to this, the fourth and final year of IMP. IMP students have performed well in SAT scores against their peers in traditional programs. Colleges and universities accept IMP as a college preparatory mathematics sequence. I know that your efforts will pay off, and I encourage you to take charge of your future by pursuing advanced math and science skills. Even if you don't become an aerospace engineer or computer programmer, this country needs people who think logically and critically, and who are well prepared to solve the issues yet to be discovered.

Larry Gilliam
Scotts Valley, California

Larry Gilliam is a parent of two IMP students and works as the chief test engineer for Lockheed Martin Missiles & Space in Sunnyvale, California.

As the Cube Turns

The Interactive Mathematics Program

What is the Interactive Mathematics Program?

The Interactive Mathematics Program (IMP) is a growing collaboration of mathematicians, teacher-educators, and teachers who have been working together since 1989 on both curriculum development and professional development for teachers.

What is the IMP curriculum?

IMP has created a four-year program of problem-based mathematics that replaces the traditional Algebra I–Geometry–Algebra II/Trigonometry–Precalculus sequence and that is designed to exemplify the curriculum reform called for in the *Curriculum and Evaluation Standards* of the National Council of Teachers of Mathematics (NCTM).

The IMP curriculum integrates traditional material with additional topics recommended by the NCTM *Standards,* such as statistics, probability, curve fitting, and matrix algebra. Although every IMP unit has a specific mathematical focus, most units are structured around a central problem and bring in other topics as needed to solve that problem, rather than narrowly restricting the mathematical content. Ideas that are developed in one unit are generally revisited and deepened in one or more later units.

For which students is the IMP curriculum intended?

The IMP curriculum is for all students. One of IMP's goals is to make the learning of a core mathematics curriculum accessible to everyone. Toward that end, we have designed the program for use with heterogeneous classes. We provide you with a varied collection of supplemental problems to give you the flexibility to meet individual student needs.

Dan Johnson and Susan Ford use a diagram as they consider a diver's initial velocity at the moment of release from the Ferris wheel's platform.

How is the IMP classroom different?

When you use the IMP curriculum, your role changes from "imparter of knowledge" to observer and facilitator. You ask challenging questions. You do not give all the answers; rather, you prod students to do their own thinking, to make generalizations, and to go beyond the immediate problem by asking themselves "What if?" The IMP curriculum gives students many opportunities to write about their mathematical thinking, to reflect on what they have done, and to make oral presentations to one another about their work. In IMP, your assessment of students becomes integrated with learning, and you evaluate students according to a variety of criteria, including class participation, daily homework assignments, Problems of the Week, portfolios, and unit assessments. The *Teaching Handbook for the Interactive Mathematics Program* provides many practical suggestions on how to get the best possible results using this curriculum in *your* classroom.

What is in Year 4 of the IMP curriculum?

Year 4 of the IMP curriculum contains five units.

High Dive

The central problem of this unit involves a circus act in which a diver is dropped from a turning Ferris wheel into a tub of water carried by a moving cart. The students' task is to determine when the diver should be released from the Ferris wheel in order to land in the moving tub of water. In analyzing this problem, students extend right-triangle trigonometric functions to the circular functions, study the physics of falling objects (including separating the diver's motion into its vertical and horizontal components), and develop an algebraic expression for the time of the diver's fall in terms of his position. Along the way, students are introduced to several additional trigonometric concepts, such as polar coordinates, inverse trigonometric functions, and the Pythagorean identity.

As the Cube Turns

This unit opens with an overhead display, generated by a program on a graphing calculator. The two-dimensional display depicts the rotation of a cube in three-dimensional space. Students' central task in the unit is to learn how to write such a program, though the real focus is on the mathematics behind the program.

As the Cube Turns

Students study the fundamental geometric transformations—translations, rotations, and reflections—in both two and three dimensions and express them in terms of coordinates. The study of these transformations also provides a new setting for students to work with matrices, which they previously studied in connection with systems of linear equations (in the Year 3 unit *Meadows or Malls?*). Another complex component of students' work is the analysis of how to represent a three-dimensional object on a two-dimensional screen. As a concluding project, students work in pairs to program an animated graphic display of their own design.

Know How

This unit is designed to prepare students to find out independently about mathematical content they either have not learned or have forgotten. Most will need this skill in later education as well as in their adult work lives. Students are given experiences of learning through reading traditional textbooks and interviewing other people. The content explored this way includes radian measure, ellipses, proof of the quadratic formula, the laws of sines and cosines, and complex numbers.

The World of Functions

This unit builds on students' extensive previous work with functions. Students explore basic families of functions in terms of various ways they can be represented—as tables, as graphs, as algebraic expressions, and as models for real-world situations. Students also use functions to explore a variety of problem situations and discover that finding an appropriate function to use as a model sometimes involves recognizing a pattern in the data and other times requires insight into the situation itself. In the last portion of the unit, students explore ways of combining and transforming functions.

The Pollster's Dilemma

The central limit theorem is the cornerstone of this unit in which students look at the process of sampling, with a special focus on how the size of the sample affects variation in poll results. The opening problem concerns an election poll, which shows 53% of the voters favoring a particular candidate.

Students investigate this question: How confident should the candidate be about her lead, based on this poll? By analyzing specific cases, students see that the results from a set of polls of a given size are approximately normally distributed. They are given the statement of the central limit theorem, which confirms this observation. Building on work in earlier units, students learn how to use normal distributions and standard deviations to find confidence intervals and see how concepts such as margin of error are used in reporting polling results. Students finish the unit by working in pairs on a sampling project for a question of their own.

How do the four years of the IMP curriculum fit together?

The four years of the IMP curriculum form an integrated sequence through which students can learn the mathematics they will need both for further education and on the job. Although the organization of the IMP curriculum is very different from the traditional Algebra I–Geometry–Algebra II/Trigonometry–Precalculus sequence, the important mathematical ideas are all there.

Here are some examples of how both traditional concepts and topics new to the high school curriculum are developed.

Solving equations

In Year 1 of the IMP curriculum, students develop an intuitive foundation of algebraic thinking, including the use of variables, which they build on throughout the program. In the Year 2 unit *Solve It!*, students use the concept of equivalent equations to see how to solve any linear equation in a single variable. In *Cookies* (Year 2) and *Meadows or Malls?* (Year 3), they solve pairs of linear equations in two or more variables, using both algebraic and geometric methods, and see how to use matrices and the technology of graphing calculators to solve such systems. In *Fireworks* (Year 3), they explore a variety of methods for solving quadratic equations, including graphing and completing the square. In *High Dive* and *Know How* (Year 4), students use the quadratic formula to solve an equation that arises from the study of falling objects, and they prove the general formula using ideas from *Fireworks*.

Geometry in One, Two, and Three Dimensions

Measurement, including area and volume, is one of the fundamental topics in geometry. In Year 1, students use angle and line measurement and their relationship to similarity to explore ideas about lengths in the unit *Shadows*. In *Do Bees*

Build It Best? (Year 2), students discover and prove the Pythagorean theorem and develop important ideas about area, volume, and surface area. Students combine these ideas with their understanding of similarity to see why the hexagonal prism of the bees' honeycomb design is the most efficient regular prism possible. In the Year 3 unit *Orchard Hideout,* students examine why the special number π appears in the formulas for both area and circumference of circles. They use these formulas, together with principles of coordinate geometry, to predict how long it will take for the center of an orchard to become a "hideout." Later in Year 3, in *Meadows or Malls?,* students extend ideas of coordinate graphing from two dimensions to three, and apply key ideas about three-dimensional graphs to solve a land-use problem. Work in three dimensions continues in Year 4, especially in *As the Cube Turns,* in which students examine how to represent a three-dimensional figure mathematically using a two-dimensional diagram.

Trigonometric functions

In traditional programs, the trigonometric functions are introduced in the eleventh or twelfth grade. In the IMP curriculum, students begin working with trigonometry in Year 1 in the unit *Shadows* and use right-triangle trigonometry in several units in Years 2 and 3, including the unit *Do Bees Build It Best?* In the Year 4 unit *High Dive,* they extend trigonometry from right triangles to circular functions in the context of a circus act in which a performer is dropped from a Ferris wheel into a moving tub of water.

Standard deviation and the binomial distribution

Standard deviation and the binomial distribution are major tools in the study of probability and statistics. The Year 1 unit *The Game of Pig* gets students started by building a firm understanding of concepts of probability and the phenomenon of experimental variation. Later in Year 1 (in *The Pit and the Pendulum*), they use standard deviation to see that the period of a pendulum is determined primarily by its length. In Year 2, students compare standard deviation with the chi-square test in examining whether the difference between two sets of data is statistically significant. In *Pennant Fever* (Year 3), students use the binomial distribution to evaluate a team's chances of winning the baseball championship, and in *The Pollster's Dilemma* (Year 4), students tie many of these ideas together in the central limit theorem, seeing how the margin of error and the level of certainty for an election poll depend on the size of the sample.

Does the program work?

The IMP curriculum has been thoroughly field-tested and enthusiastically received by hundreds of classroom teachers around the country. Their enthusiasm is based on the success they have seen in their own classrooms with their own students. For instance, IMP teacher Dennis Cavaillé says, "For the first time in my teaching career, I see lots of students excited about solving math problems inside *and* outside of class."

These informal observations are backed up by more formal evaluations. Dr. Norman Webb of the Wisconsin Center for Education Research has done several studies comparing the performance of students using the IMP curriculum with the performance of students in traditional programs. For instance, he has found that IMP students do as well as students in traditional mathematics classes on standardized tests such as the SAT. This is especially significant because IMP students spend about twenty-five percent of their time studying topics, such as statistics, not covered on these tests. To measure IMP students' achievement in these other areas, Dr. Webb conducted three separate studies involving students at different grade levels and in different locations. The three tests used in these studies involved statistics, quantitative reasoning, and general problem solving. In all three cases, the IMP students outperformed their counterparts in traditional programs by a statistically significant margin, even though the two groups began with equivalent scores on eighth grade standardized tests.

But one of our proudest achievements is that IMP students are excited about mathematics, as shown by Dr. Webb's finding that they take more mathematics courses in high school than their counterparts in traditional programs. We think this is because they see that mathematics can be relevant to their own lives. If so, then the program works.

Dan Fendel

Diane Resek

Lynne Alper

Sherry Fraser

As the Cube Turns

Note to Students

This textbook represents the last year of a four-year program of mathematics learning and investigation. As in the first three years, the program is organized around interesting, complex problems, and the concepts you learn grow out of what you'll need to solve those problems.

If you studied IMP Year 1, 2, or 3

If you studied IMP Year 1, 2, or 3, then you know the excitement of problem-based mathematical study. The Year 4 program extends and expands the challenges that you worked with previously. For instance:

- In Year 1, you began developing a foundation for working with variables. In Year 2, you learned how to solve linear equations algebraically, and in Year 3, you worked with quadratic equations. In Year 4, you'll solve a quadratic equation as part of the process of finding out when a diver should be dropped from a Ferris wheel in order to land in a moving tub of water.

- In Year 1, you used the normal distribution to help predict the period of a 30-foot pendulum. In Year 2, you learned about the chi-square statistic to understand statistical comparisons of populations, and in Year 3, you learned about the binomial distribution. In Year 4, you'll use the context of election polls to see the connection between the binomial distribution and the normal distribution, and you'll use ideas such as margin of error and confidence level to study how sample size affects poll reliability.

You'll also use ideas from geometry, trigonometry, and matrix algebra to develop a calculator program that shows a cube rotating in space, you'll prove the quadratic formula as part of a unit on ways to learn mathematics on your own, and you'll synthesize your IMP experience with functions by examining a variety of methods for creating functions that fit specific real-world problems.

These pages in the student book welcome students to the program.

Interactive Mathematics Program

Welcome! Year 4

If you didn't study IMP Year 1, 2, or 3
If this is your first experience with the Interactive Mathematics Program (IMP), you can rely on your classmates and your teacher to fill in what you've missed. Meanwhile, here are some things you should know about the program, how it was developed, and how it is organized.

The Interactive Mathematics Program is the product of a collaboration of teachers, teacher-educators, and mathematicians who have been working together since 1989 to reform the way high school mathematics is taught. About one hundred thousand students and five hundred teachers used these materials before they were published. Their experiences, reactions, and ideas have been incorporated into this final version.

Our goal is to give you the mathematics you need in order to succeed in this changing world. We want to present mathematics to you in a manner that reflects how mathematics is used and that reflects the different ways people work and learn together. Through this perspective on mathematics, you will be prepared both for continued study of mathematics in college and for the world of work.

This book contains the various assignments that will be your work during Year 4 of the program. As you will see, these problems require ideas from many branches of mathematics, including algebra, geometry, probability, graphing, statistics, and trigonometry. Rather than present each of these areas separately, we have integrated them and presented them in meaningful contexts, so you will see how they relate to each other and to our world.

Each unit in this four-year program has a central problem or theme, and focuses on several major mathematical ideas. Within each unit, the material is organized for teaching purposes into "days," with a homework assignment for each day. (Your class may not follow this schedule exactly, especially if it doesn't meet every day.)

At the end of the main material for each unit, you will find a set of supplementary problems. These problems provide you with additional opportunities to work with ideas from the unit, either to strengthen your understanding of the core material or to explore new ideas related to the unit.

Although the IMP program is not organized into courses called "Algebra," "Geometry," and so on, you will be learning all the essential mathematical concepts that are part of those traditional courses. You will also be learning concepts from branches of mathematics—especially statistics and probability—that are not part of a traditional high school program.

To accomplish your goals, you will have to be an active learner, because the book does not teach directly. Your role as a mathematics student will be to experiment, to investigate, to ask questions, to make and test conjectures, and to reflect, and then to communicate your ideas and conclusions both orally and in writing. You will do some of your work in collaboration with fellow students, just as users of mathematics in the real world often work in teams. At other times, you will be working on your own.

We hope you will enjoy the challenge of this new way of learning mathematics and will see mathematics in a new light.

Dan Fendel   Diane Resek

Lynne Alper   Sherry Fraser

As the Cube Turns

Finding What You Need

We designed this guide to help you find what you need amid all the information it provides. Each of the following components has a special treatment in the layout of the guide.

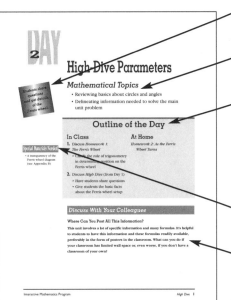

- **Synopsis of the Day:** The key idea or activity for each day is summarized in a brief sentence or two.

- **Mathematical Topics:** Mathematical issues for the day are presented in a bulleted list.

- **Outline of the Day:** Under the *In Class* heading, the outline summarizes the activities for the day, which are keyed to numbered headings in the discussion. Daily homework assignments and Problems of the Week are listed under the *At Home* heading.

- **Special Materials Needed:** Special items needed in the classroom for each day are bulleted here.

- **Discuss With Your Colleagues:** This section highlights topics that you may want to discuss with your peers.

- **Post This:** The *Post This* icon indicates items that you may want to display in the classroom.

- **Suggested Questions:** These are specific questions that you might ask during an activity or discussion to promote student insight or to determine whether students understand an idea. The appropriateness of these questions generally depends on what students have already developed or presented on their own.

- **Asides:** These are ideas outside the main thrust of a discussion. They include background information, refinements or subtle points that may only be of interest to some students, ways to help fill in gaps in understanding the main ideas, and suggestions about when to bring in a particular concept.

Icons for Student Written Products

Single group report   **Individual reports**

As the Cube Turns

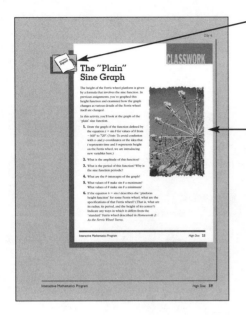

Icons for Student Written Products: For each group activity, there is an icon suggesting a single group report, individual reports, or no report at all. If graphs are included, the icon indicates this as well. (The graph icons do not appear in every unit.)

Embedded Student Pages: The teacher guide contains reduced-size copies of the pages from the student book, including the "transition pages" that appear occasionally within each unit to summarize each portion of the unit and to prepare students for what is coming. The reduced-size classwork and homework assignments follow the teacher notes for the day on which the activity is begun. Having all of these student pages in the teacher's guide is a helpful way for you to see things from the students' perspective.

Additional Information

Here is a brief outline of other tools we have included to assist you and make both the teaching and the learning experience more rewarding.

Glossary: This section, which is found at the back of the book, gives the definitions of important terms for all of Year 4 for easy reference. The same glossary appears in the student book.

Appendix A: Supplemental Problems: This appendix contains a variety of interesting additional activities for the unit, for teachers who would like to supplement material found in the regular classroom problems. These additional activities are of two types—*reinforcements,* which help increase student understanding of concepts that are central to the unit, and *extensions,* which allow students to explore ideas beyond the basic unit.

Appendix B and Appendix C: Blackline Masters: For each unit, these appendices contain materials you can reproduce that are not available in the student book and that will be helpful to teacher and student alike. They include the end-of-unit assessments as well as such items as diagrams from which you can make transparencies. Semester assessments for Year 4 are included in *As the Cube Turns* (for first semester) and *The Pollster's Dilemma* (for second semester).

Single group graph

Individual graphs

No report at all

Interactive Mathematics Program

Year 4 IMP Units

High Dive

As the Cube Turns (in this book)

Know How

The World of Functions

The Pollster's Dilemma

CONTENTS

As the Cube Turns Overview xxviii
   Summary of the Unit xxviii
   Concepts and Skills xxix
   Materials xxx
   Grading xxxi
   Days 1-2: Picture This! (Reduced Student Page) 2

Day 1: Picture This! 3
   Forming Groups 4
   Introducing the Unit 4
   Picture This! 7
   POW 3: "A Sticky Gum Problem" Revisited 9
   Homework 1: Starting Sticky Gum 9

Day 2: Continuing to *Picture This!* 15
   Discussion of *Homework 1: Starting Sticky Gum* 16
   Continued Work on *Picture This!* 16
   Writing a Program 17
   Homework 2: Programming Without a Calculator 18
   Days 3-7: Programming Loops (Reduced Student Page) 21

Day 3: Around and Around We Go! 23
   Discussion of *Homework 2: Programming Without a Calculator* ... 23
   The Basics of Loops 25
   Using the For/End Variable 28
   Homework 3: Learning the Loops 30

Day 4: Animation 33
   Discussion of *Homework 3: Learning the Loops* 33
   Where Are We in the Unit? 35
   Flip Books: The Illusion of Motion 36

Interactive Mathematics Program

An Animated Shape 36

Homework 4: A Flip Book 37

Day 5: Busy Work 41

Discussion of *Homework 4: A Flip Book* 41

Completion and Discussion of *An Animated Shape* 42

The Delay 42

Homework 5: Movin' On 43

Day 6: Set It Up! 45

POW Presentation Preparation 45

Discussion of *Homework 5: Movin' On* 46

More Presentations from *An Animated Shape* 48

Homework 6: Some Back and Forth 48

Day 7: POW 3 Presentations and a Moving Arrow 51

Presentations of *POW 3: "A Sticky Gum Problem"* Revisited 51

Discussion of *Homework 6: Some Back and Forth* 52

Arrow 52

Homework 7: Sunrise 52

POW 4: A Wider Windshield Wiper, Please 52

Days 8–10: Translation in Two Dimensions (Reduced Student Page) 57

Day 8: Move That Line! 59

Discussion of *Homework 7: Sunrise* 59

Introduction of Geometric Transformations 60

Move That Line! 63

Homework 8: Double Dotting 65

Day 9: Continuing to Move the Line 69

Discussion of *Homework 8: Double Dotting* 69

Completion of *Move That Line!* 70

Discussion of *Move That Line!* 70

Homework 9: Memories of Matrices 73

Day 10: A Matrix Day 77

Discussion of *Homework 9: Memories of Matrices* 78

Moving the Line with Matrices 79

Homework 10: Cornering the Cabbage 80

Days 11-19: Rotating in Two Dimensions (Reduced Student Page) 83

Day 11: An Area Formula 85

Discussion of *Homework 10: Cornering the Cabbage* 86

Where Are We in the Unit? 88

Rotations: An Introduction 89

Homework 11: Goin' Round the Origin 89

Day 12: Goin' Round the Origin 91

Discussion of *Homework 11: Goin' Round the Origin* 92

Review of Polar Coordinates 93

Homework 12: Double Trouble 95

Day 13: The Sine of a Sum 99

Forming New Groups 99

Discussion of *Homework 12: Double Trouble* 99

The Sine of a Sum 101

Homework 13: A Broken Button 101

Day 14: Conclusion of *The Sine of a Sum* 105

Discussion of *Homework 13: A Broken Button* 105

Discussion of *The Sine of a Sum* 107

Homework 14: Oh, Say What You Can See 109

Day 15: The Cosine of a Sum 113

Discussion of *Homework 14: Oh, Say What You Can See* 113

The Cosine of a Sum 114

Rotations Revisited 117

Homework 15: Comin' Round Again (and Again ...) 118

Day 16: Rotation with Matrices 121

Discussion of *Homework 15: Comin' Round Again (and Again ...)* 121

More Memories of Matrices 122

Homework 16: Taking Steps 122

Day 17: Rotation with Matrices, Continued 127
Discussion of *Homework 16: Taking Steps* 127
Discussion of *More Memories of Matrices* 128
Homework 17: How Did We Get Here? 131

Day 18: Swing That Line! 133
Discussion of *Homework 17: How Did We Get Here?* 134
Swing That Line! 134
Homework 18: Doubles and Differences 136

Day 19: Swing That Line Some More! 139
POW Presentation Preparation 139
Discussion of *Homework 18: Doubles and Differences* 139
Conclusion and Discussion of *Swing That Line!* 140
Homework 19: What's Going On Here? 141
Days 20-30: Projecting Pictures (Reduced Student Page) 143

Day 20: POW 4 Presentations 145
Discussion of *Homework 19: What's Going On Here?* 146
Introduction to Projections 147
Presentations of *POW 4: A Wider Windshield Wiper, Please* 148
Introducing *POW 5: An Animated POW* 149
Homework 20: "A Snack in the Middle" Revisited 150

Day 21: Orchard Snacks 153
POW Partner Reminder 153
Discussion of *Homework 20: "A Snack in the Middle" Revisited* 154
Fractional Snacks 156
Discussion of *Fractional Snacks* 156
Homework 21: More Walking for Clyde 158

Day 22: Monorail Delivery 161
Collect POW 5 Partner Names 161
Discussion of *Homework 21: More Walking for Clyde* 161
Monorail Delivery 162

As the Cube Turns

Discussion of *Monorail Delivery* 163
Homework 22: *Another Mystery* 164

Day 23: Return to the Third Dimension 167

Forming New Groups 168
Reminder on POW Descriptions 168
Discussion of *Homework 22: Another Mystery* 168
Reviewing the Three-Dimensional Coordinate System 169
A Return to the Third Dimension 171
Homework 23: *Where's Madie?* 171

Day 24: And Fred Brings the Lunch 175

Collect Descriptions for POW 175
Discussion of *Homework 23: Where's Madie?* 175
And Fred Brings the Lunch 176
Homework 24: *Flipping Points* 176

Day 25: Fred Brings More Lunch 181

Discussion of *Homework 24: Flipping Points* 181
Discussion of *And Fred Brings the Lunch* 182
Homework 25: *Where's Bonita?* 183

Day 26: Lunch in the Window 185

Discussion of *Homework 25: Where's Bonita?* 185
Lunch in the Window 186
Discussion of *Lunch in the Window* 186
Homework 26: *Further Flips* 188

Day 27: Cube on a Screen 193

Discussion of *Homework 26: Further Flips* 193
Cube on a Screen 194
Discussion of *Cube on a Screen* 196
Projection Using Coordinates: An Overview 197
Homework 27: *Spiders and Cubes* 197

Day 28: Find Those Corners! 203

Reminders on *POW 5: An Animated POW* 203
Discussion of *Homework 27: Spiders and Cubes* 204

Find Those Corners! 204

Homework 28: *An Animated Outline* 209

Day 29: Finding More Corners 213

Collection of Homework 28: *An Animated Outline* 213

Continued Work on *Find Those Corners!* 213

Homework 29: *Mirrors in Space* 214

Day 30: Finishing the Corner 217

Discussion of Homework 29: *Mirrors in Space* 217

Discussion of *Find Those Corners!* 218

How to Project 219

Homework 30: *Where Are We Now?* 219

Days 31–33: Rotating in Three Dimensions (Reduced Student Page) ... 221

Day 31: Rotating Around It 223

Discussion of Homework 30: *Where Are We Now?* 223

Rotating Around the z-axis 224

Follow That Point! 225

Discussion of *Follow That Point!* 225

Homework 31: *One Turn of a Cube* 226

Day 32: Rotating Matrices 229

Discussion of Homework 31: *One Turn of a Cube* 229

Rotation Matrix in Three Dimensions 230

Discussion of Rotation Matrix in Three Dimensions 230

Homework 32: *The Turning Cube Outline* 231

Day 33: *As the Cube Turns* 235

Discussion of Homework 32: *The Turning Cube Outline* 235

Concluding *As the Cube Turns* 236

Homework 33: *Beginning Portfolio Selection* 237

Days 34–36: An Animated POW (Reduced Student Page) 239

Day 34: Completely Animated 241

Discussion of Homework 33: *Beginning Portfolio Selection* 241

Completion of *POW 5: An Animated POW* 242

As the Cube Turns

    Optional: Animation Video 242

    Homework 34: "An Animated POW" Write-up 243

Day 35: POW 5 Presentations 245

    Presentations of *POW 5: An Animated POW* 245

    Homework 35: Continued Portfolio Selection 245

Day 36: Finishing Presentations 247

    Discussion of *Homework 35: Continued Portfolio Selection* 247

    Continued Presentations on *POW 5: An Animated POW* 247

    Homework 36: "As the Cube Turns" Portfolio 248

Day 37: Final Assessments 251

    End-of-Unit Assessments 251

Day 38: Summing Up 255

    Discussion of Unit Assessments 255

    Unit Summary 257

Appendix A: Supplemental Problems 259

    Appendix: Supplemental Problems (Reduced Student Page) 261

    Loopy Arithmetic 262

    Sum Tangents 264

    Moving to the Second Quadrant 265

    Adding 180° 267

    Sums for All Quadrants 268

    Bugs in Trees 269

    Half a Sine 273

    The General Isometry 274

    Perspective on Geometry 276

    Let the Calculator Do It! 277

Appendix B: TURNCUBE 279

Appendix C: Blackline Masters 283

Glossary 295

As the Cube Turns Overview

Summary of the Unit

This unit opens with an overhead display, generated by a program on a graphing calculator, that shows a cube rotating in three-dimensional space. The central problem of the unit is for students to learn what goes into writing such a program.

In addition to introducing students to programming issues such as the use of loops, this task takes them into several areas of mathematics. They study the fundamental geometric transformations—translations, rotations, and reflections—in two dimensions and express them in terms of coordinates.

Students also need to learn about rotations in three dimensions in order to work on the unit problem. The analysis of rotations builds on students' experience (in the Year 4 unit *High Dive*) with trigonometric functions and polar coordinates, and leads them to see the need for and to develop formulas for the sine and cosine of the sum of two angles. The study of these geometric transformations also provides a new setting for students to work with matrices, which they previously studied in connection with systems of linear equations (in the Year 3 unit *Meadows or Malls?*).

Another important component of students' work is the analysis of how to represent a three-dimensional object on a two-dimensional screen. This problem is approached through the use of both physical materials and an amusing story about two spiders and a thread of web that connects them. Students see how projection onto a plane is affected by both the choice of the plane and the choice of a viewpoint or center of projection.

The unit closes with a project in which students program an animated graphic display of their own design.

The outline below gives a summary of the unit's overall organization in terms of the daily schedule.

- Days 1-2: Exploring the graphing calculator's capacity for drawing pictures

- Days 3-7: Learning about programming loops, using a For/End type of instruction combination

- Days 8-10: Expressing two-dimensional translations in terms of coordinates and matrices

- Days 11-19: Developing trigonometric formulas and using them to express two-dimensional rotations in terms of coordinates and matrices

As the Cube Turns

- Days 20–30: Developing the geometric ideas behind projection from a viewpoint onto a plane and expressing projection in terms of coordinates
- Days 31–33: Expressing three-dimensional rotations in terms of coordinates and matrices and completing the unit problem
- Days 34–36: Completing the animation projects and presenting them to the class
- Days 37–38: Assessments and summing up

Concepts and Skills

Here is a summary of the main concepts and skills that students will encounter and practice in this unit.

Coordinate geometry

- Expressing geometric transformations—translations, rotations, and reflections—in terms of coordinates in two and three dimensions
- Finding coordinates a fractional distance along a line segment in two and three dimensions
- Reviewing graphing in three dimensions
- Finding the projection of a point onto a plane from the perspective of a fixed point and developing an algebraic description of the projection process
- Studying the effect of change of viewpoint on projections
- Reviewing polar coordinates

Matrices

- Reviewing the algebra of matrices
- Using matrices to express geometric transformations in two and three dimensions

Programming

- Learning to use a technical manual
- Using loops in programming
- Understanding programs from their listings
- Designing and programming animation

Synthetic geometry and trigonometry

- Reviewing formulas relating the sine of an angle to the cosine of a related angle

Interactive Mathematics Program

- Deriving the formula for the area of a triangle in terms of the lengths of two sides and the sine of the included angle
- Deriving formulas for the sine and cosine of the negative of an angle
- Deriving formulas for the sine and the cosine of the sum of two angles

Materials

You will need to provide these materials during the course of the unit (in addition to standard materials such as graphing calculators, transparencies, chart paper, and marking pens).

- A calculator connected to an overhead display (for use throughout the unit)
- A program to show the turning cube (The program called "TURNCUBE" for the TI-82 calculator is given in Appendix B and is also available on the publisher's Web site.)
- A large styrofoam cube (or similar object) and two sticks for a classroom demonstration (on Day 1 and perhaps again on Day 31)
- (Optional) A centimeter cube and two straws (or similar objects) for each group
- A calculator manual for each pair of students (for use throughout the unit, especially on Days 1, 2, 10, and 34–36)
- (Optional) A box and slips of paper for students to "act out" programming loops
- Six to eight index cards per student for *Homework 4: A Flip Book*
- (Optional) A sample flip book
- Sheets of Plexiglas or some other firm, clear material (one for Day 20 and one per pair of students for Day 27)
- Cubes, measuring at least 2 inches on each edge (one for Day 20 and one per pair of students for Day 27). If possible, these cubes should be made by connecting smaller cubes of different colors at the corners.
- (Optional) String for making a large-size three-dimensional coordinate system, with its origin in the center of the classroom, on Day 23
- Three large index cards or pieces of tag board, 5" × 8" or larger, per group (for making a model of the three-dimensional coordinate system on Day 23)
- Yarn and a pair of scissors for each group (for Day 23)

- Pens of three different colors for writing on the firm, clear material on Day 27 (a set of pens for each pair of students)
- (Optional) A video on the making of animation features, such as *The Making of Toy Story* (for use as a unit opener or closer or at any convenient time in the unit)
- (Optional) Software to connect a graphing calculator to a computer for writing, editing, and storing calculator programs.

Appendix C contains diagrams from which you can create transparencies for use in various discussions. It also contains the in-class and take-home assessments, which you will need to reproduce for Day 37.

Technology note

This unit was initially developed with the Texas Instruments TI-82 calculator in mind, and the activities work just as well with the TI-83 calculator. We recommend that before using the unit with any other graphing calculator, you confirm that the necessary matrix work and screen-drawing instructions can be carried out on that calculator. (The TI-81 calculator is not sufficient for the work in this unit.)

If you are using one of the Texas Instruments calculators, we recommend that you obtain a copy of the IMP Year 4: *Calculator Guide for the TI-82 and TI-83,* which provides details on how the programming concepts and other calculator ideas used in this unit are carried out on the TI-82 or TI-83, as well as further suggestions on calculator use in this unit. (This guide is available from Key Curriculum Press at no cost with the purchase of a class set of student textbooks. Similar guides are available for IMP Years 1, 2, and 3.)

Grading

The IMP *Teaching Handbook* contains general guidelines about how to grade students in an IMP class. You will probably want to check daily that students have done their homework and include regularity of homework completion as part of students' grades. Your grading scheme will probably also include Problems of the Week, the unit portfolio, and the end-of-unit assessments.

Because you will not be able to read thoroughly every assignment that students turn in, you will need to select certain assignments to read carefully and to base grades on. Here are some suggestions.

- *Homework 3: Learning the Loops*
- *Move That Line!* (completed on Day 9)
- *Homework 14: Oh, Say What You Can See*
- *Swing That Line!* (Day 19)

- *And Fred Brings the Lunch* (Day 25)
- *Find Those Corners!* (completed on Day 30)
- Work on POW5: *An Animated POW* (The outline is turned in on Day 33; the write-up is turned in on Day 35, and presentations are made on Days 35 and 36.)

If you want to base your grading on more tasks, there are many other homework assignments, class activities, and oral presentations you can use.

Interactive Mathematics Program®

Integrated High School Mathematics

YEAR 4

As the Cube Turns

Day 1

DAYS 1-2

Picture This!

This page in the student book introduces Days 1 and 2.

Stephanie Wood and Elizabeth Graf begin the unit by conferring about how to draw pictures on the graphing calculator.

Did you ever wonder how animators create all of those wonderful effects you see on the movie screen? In this unit, you will learn the mathematics behind computer animation and create your own animation programs on a graphing calculator.

You begin this unit by discovering how to use your calculator to draw pictures.

DAY 1

Picture This!

Students see the cube turning on the calculator screen and begin to learn calculator drawing techniques.

Mathematical Topics

- Introducing the central problem of the unit
- Learning to use a technical manual
- Learning to draw on a graphing calculator

Outline of the Day

In Class

1. Form random groups
2. Introduce the central task of the unit
   - Have students view a calculator program that shows a revolving cube, and give a classroom simulation
   - Describe the unit task
   - Discuss the overall plan for the unit
3. *Picture This!*
   - Students work with a manual to discover how to draw on their calculator
   - Students take notes and share their discoveries with the class as they make them
   - Work on this activity will be completed tomorrow

At Home

Homework 1: Starting Sticky Gum

POW 3: "A Sticky Gum Problem" Revisited (due Day 7)

Special Materials Needed

- A calculator connected to an overhead projector (for use throughout the unit)
- The program TURNCUBE (see Appendix B), or a similar program, entered in the calculator
- A large Styrofoam cube (or similar object) and two sticks for a classroom demonstration
- A calculator manual for each pair of students
- (Optional) A centimeter cube and two straws for each group
- (Optional) A video on the making of animation features

Interactive Mathematics Program

As the Cube Turns 3

Day 1

Discuss With Your Colleagues

How Do I Deal with These Special Materials?

This unit involves some unusual materials, such as the Styrofoam cube today or the Plexiglas sheets (or other firm clear material) for the activity *Cube on a Screen* on Day 27. As a mathematics teacher, you are probably not accustomed to having to round up such materials, and your department may not have a budget for it. Discuss ways to manage the logistics and the finances of that activity and others.

Note: The overhead display for the calculator will be helpful throughout the unit and is not specifically listed on individual days. Calculator manuals, while useful all the time, are especially needed on Days 1, 2, and 10. Students may also want to refer to them as they work on their animation projects, especially when they complete their work on Day 34. Manuals are included in the materials lists for those specific days.

1. Forming Groups

At the beginning of the unit, group students as described in the IMP *Teaching Handbook,* recording the members of each group and the suit for each student. We recommend that new groups be formed again on Day 13 and on Day 23.

2. Introducing the Unit

> *Note:* If you have trigonometric identities still posted from the *High Dive* unit, we suggest that you keep them up. If wall space is scarce, you can put them on top of one another, under another poster from this unit.

Run the program called TURNCUBE (or a similar program) on your graphing calculator overhead projector. (The program shows a cube, situated in 3-space, making one complete rotation around the z-axis, as described further in this section. The actual TURNCUBE program, which was written for the Texas Instruments TI-82 calculator, is given in Appendix B. The program is also available on the Key Curriculum Press World Wide Web page at http://www.keypress.com.)

Discuss with the class what physical motion of the cube the program is supposed to be representing on the screen, and demonstrate with a concrete model of some kind. (Any sort of box, and not necessarily a perfect cube, will do.) We suggest that you set this up using a classroom wall as the xy-plane, with the z-axis perpendicular to that plane. The cube should go around the z-axis at a constant distance from the wall.

Day 1

As illustrated in the diagram, you might represent the z-axis by a stick perpendicular to the wall and use another stick, stuck into a box (such as a Styrofoam cube), to demonstrate what's happening. The second stick will connect the box to the z-axis, with one end of that stick stuck into the box and the other end at and perpendicular to the z-axis. The second stick then turns with the end at the z-axis acting as a pivot point, so that the box at the other end gets turned around the axis.

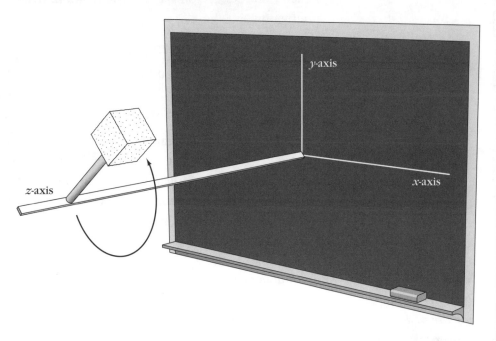

Bring out that as the box makes one complete turn around the z-axis, it also rotates once. For instance, suppose the box starts with its "bottom" horizontal (roughly in the xz-plane). Then, after the box makes a quarter turn around the z-axis, the face that was on the bottom will be vertical (roughly in the yz-plane).

You may want to give each group a cube with a hole in it and two straws so that students can re-create the demonstration for themselves. They should draw the x- and y-axes on a piece of paper, hold one straw at the origin, perpendicular to the xy-plane, and place the second straw with one end at and perpendicular to the z-axis and the other end attached to the cube.

"Is the motion of the cube more like that of the moon about the earth or the earth about the sun?"

You can ask students whether the motion of the cube is more like that of the moon about the earth or the earth about the sun. If no one knows the difference, you can tell them that the earth rotates as it orbits the sun, while the moon always keeps the same face toward the earth. Thus, the cube moves the same way the moon does.

After the demonstration, run the calculator program again for students, now that they better understand what the program is supposed to be portraying. Tell them that over the course of the unit, they will be learning the programming skills and mathematics concepts needed to write a program

Day 1

that creates the image of this moving cube on the calculator screen. You might note that this may not seem like a very exciting animation, but its program will be surprisingly complex. Students will learn techniques that are basic to all computer animation.

Tell students that as an end-of-the-unit project, they will work in pairs to create original animation programs. (This is *POW 5: An Animated POW,* which is introduced on Day 20. You might have students look at this briefly now.)

• *Optional: An animation video*

At some point during the unit, you may want to show a video such as *The Making of Toy Story,* which describes the making of a major animation feature film. A local television station may be a source for such videos.

This idea is mentioned again near the end of the unit (see Day 34). If you show a video like this at the end of the unit, students will see how what they have learned is applied in real-world animation. If you show it at the beginning of the unit, students may get additional motivation. In fact, a video like this can be used whenever convenient.

• *The plan of the unit*

"What do you think you will need to learn in order to write a program to make the cube turn?"

Ask students what they think they will need to learn in this unit in order to write a program to show the cube turning. We recommend that you build on their ideas to develop this outline of the unit.

Students will learn to do these things.

1. Draw a picture on the graphing calculator.

2. Create the appearance of motion.

3. Change the position of an object located in a two-dimensional coordinate system.

4. Create a two-dimensional drawing of a three-dimensional object.

5. Change the position of an object located in a three-dimensional coordinate system.

Some teachers have found it very useful to have this plan posted for reference throughout the unit. It serves as an organizer so that students can see what they have learned and what they are going to be learning.

Day 1

> *Comment:* Items 3 and 5 involve the geometrical transformations called *translations* and *rotations*. (The geometric meaning of the term *translation* is introduced on Day 8.)
>
> Strictly speaking, to accomplish the unit task of creating a program for animating a cube, students only need rotations in three dimensions. But translations are much easier to understand and work with, and it's easier to begin with the two-dimensional case, so that's how the unit is organized.
>
> In fact, students do not ever discuss translations in three dimensions, though we suggest that you mention on Day 11 that translations in three dimensions are very similar to translations in two dimensions.
>
> The word *projection* (which is the formal term for the task in item 4) is introduced on Day 20.

Tell students that not only will they learn the coordinate geometry and algebra involved in these tasks, but they will write programs to show these things on the graphing calculator.

3. *Picture This!*

> Students' first task in this unit is to learn how to make a drawing on the graphing calculator. In the opening activity, *Picture This!,* students explore the drawing features of their calculators. This activity goes well if students work in pairs.
>
> Students likely will focus today on working in a "direct mode" in which they give the calculator instructions to be carried out immediately. Tomorrow, they will learn about putting drawing instructions into a program.
>
> One of the goals of this activity—and the unit—is for students to gain experience working with a technical manual. Thus, you should make instruction manuals for the calculators available as students work on this activity.
>
> Students should be encouraged to learn from each other. They should be aware of what other pairs are doing, and they should keep careful notes of their discoveries for future reference.

Have students begin work on *Picture This!* As they work, provide opportunities for them to share discoveries with the whole class, perhaps using the calculator overhead projector. You may want to create a protocol by which pairs can interrupt the rest of the class to share an idea. Encourage other students to question the presenters to be sure that they can all replicate the results.

Emphasize to students that they will be applying what they learn in this activity throughout the unit, so they should keep careful notes. You might suggest that they keep a separate notebook (or notebook section) for programming and calculator ideas and techniques.

"What effects would you like to create? What might you try to get the effect you want?"

As you circulate among the groups, ask students to think about what effects they would like to create, and encourage them to experiment in trying to create those effects. They can simply try some guesses, or they can refer to the manual or ask other students for suggestions. Assure them that they can't harm the calculators by experimenting.

More time is allotted tomorrow for this exploratory work. Here are three things that students should know how to do by the end of the work on this activity.

- How to clear the calculator screen
- How to draw on the screen a line segment that connects two points that are given in terms of their coordinates
- How to draw on the screen a circle that is given in terms of its radius and the coordinates of its center

- *Calculator expertise*

    Many teachers do not feel completely comfortable with the programming concepts and other calculator ideas used in this unit. We have two suggestions for you.

    - Obtain a copy of *IMP Year 4: Calculator Guide for the TI-82 and TI-83* (see the "Technology note" in the Materials section of the unit overview).
    - Identify the student calculator experts. You will likely have some students who are at least as well informed about calculator usage as you, and you can have them assist their classmates in debugging programs and overcoming technical obstacles.

- *Calculators and equity*

    Students who have their own graphing calculators may have a significant advantage over students who do not. In particular, students with greater access to calculators will be able to try out programming ideas at their leisure, while other students can test ideas only during class.

    We urge you to do what you can to equalize the situation. For example, you may be able to have students check out school-owned calculators to take home, or you may be able to provide access to calculators during lunchtime or after school.

Day 1

POW 3: "A Sticky Gum Problem" Revisited

> This assignment is a variation on *POW 4: A Sticky Gum Problem* from the Year 1 unit *The Game of Pig*. It is scheduled to be discussed on Day 7.

Homework 1: Starting Sticky Gum

> Tonight's homework simply gets students started on *POW 3: "A Sticky Gum Problem" Revisited*.

Amanda Jones, Matt Solie, Briann McCoy, and Candice Anderson begin the unit by exploring how to draw pictures on the graphing calculator.

Interactive Mathematics Program   As the Cube Turns 9

Day 1

Picture This!

Your goal today is to learn about drawing pictures on your graphing calculator, which is the first step toward making a turning cube on the calculator.

You can begin simply by experimenting with different keys or menus on the calculator. You may want to read parts of the calculator instruction manual or look up ideas in its index or table of contents. You can also ask other students for ideas.

Take careful notes on your own discoveries and on ideas you get from fellow students. You may appreciate these notes later in the unit as you work on the turning cube program or on your project.

"A Sticky Gum Problem" Revisited

Do you remember Ms. Hernandez and her twins? They were the main characters in a Year 1 POW called "A Sticky Gum Problem" (in the unit *The Game of Pig*). This POW involves a variation on that POW.

The Original POW

To refresh your memory, here's the initial scenario from the Year 1 POW.

> Every time Ms. Hernandez passed a gum ball machine, her twins each wanted to get a gum ball, and they insisted on getting gum balls of the same color. Gum balls cost a penny each, and Ms. Hernandez had no control over which color she would get.

First, Ms. Hernandez and the twins passed a gum ball machine with only two colors. Then they came across a machine with three colors. In each case, you needed to find out the maximum amount Ms. Hernandez might have to spend in order to satisfy her twins.

Then Mr. Hodges came by the three-color gum ball machine. He had triplets, and so he needed to get three gum balls that were the same color. You needed to find the maximum amount Mr. Hodges might have to spend in order to satisfy his triplets.

Finally, you were asked to generalize the problem. Your goal was to find a formula that would work for any number of colors and any number of children. Your formula needed to tell you the maximum amount that the parent might have to spend to provide each of the children with a gum ball of the same color.

Continued on next page

1. Before starting on the new problem, re-create the generalization for the old one.

   a. Find a formula for the maximum amount a parent might have to spend in terms of the number of colors and the number of children.

   b. Provide a proof of your generalization. That is, give a convincing argument to show that your formula is correct.

Some New Gum Ball Problems

Ms. Hernandez' twins have grown up a bit in the last three years, and they have changed in some ways. Now they each insist on getting a gum ball of a *different* color from the other twin. (Gum balls are still a penny each.)

One day, Ms. Hernandez and the twins passed a gum ball machine that contained exactly 20 gum balls: 8 yellow, 7 red, and 5 black. As before, Ms. Hernandez could not control which gum ball would come out of the machine.

2. If each twin wanted one gum ball, what's the maximum amount that Ms. Hernandez might have to spend so that the twins could each get a different color? Prove your answer.

3. Because the twins have grown, they now have bigger mouths, so sometimes they each want two gum balls. The two gum balls each twin gets must be the same color, so the flavors match, but one twin's pair of gum balls have to be a different color from the pair the other twin gets.

What's the maximum amount that Ms. Hernandez might have to spend? Prove your answer.

Continued on next page

New Generalizations

Now, make at least three generalizations about these new problems. You get to decide exactly what you want to generalize. Here are some options.

- Generalize the number of children.
- Generalize the number of gum balls each child wants.
- Generalize the number of gum balls of each color.
- Generalize the number of colors in the machine.

Your formula or procedure should tell you how to find the maximum amount that the parent might have to spend to provide all the children with the particular number of gum balls with each child getting gum balls of a different color.

Write a proof for each generalization you create. As a grand finale, try to generalize all of these variables.

Write-up

Your write-up for this problem should begin with the formula and proof for Question 1 and the answers (with proofs) for Questions 2 and 3.

Then present each generalization you found for the new type of problem, with a proof for each generalization. Also explain how you arrived at each generalization and how you discovered your proof.

Adapted from "A Sticky Gum Problem" in *aha! Insight* by Martin Gardner, W. H. Freeman and Company, New York City/San Francisco, 1978.

HOMEWORK 1

Starting Sticky Gum

Read *POW 3: "A Sticky Gum Problem" Revisited*.

Assume that a parent comes across a gum ball machine containing many gum balls of several colors, and needs to provide each of several children with a gum ball of the same color.

1. (This is Question 1 of the POW.)

   a. Find a formula for the maximum amount a parent might have to spend in terms of the number of colors in the gum ball machine and the number of children.

   b. Provide a proof of your generalization. That is, give a convincing argument to show that your formula is correct.

Suppose Ms. Hernandez and the twins passed a gum ball machine that contained 20 gum balls: 8 yellow ones, 7 red ones, and 5 black ones.

2. (This is Question 2 of the POW.) If each twin wanted one gum ball, what's the maximum amount that Ms. Hernandez might have to spend so that the twins could each get a *different* color? Prove your answer.

3. Identify any questions you have about what is expected in the POW.

DAY 2

Continuing to *Picture This!*

Students continue to share discoveries about their graphing calculators.

Mathematical Topics

• Continuing to work with a technical manual
• Using a graphing calculator to draw figures
• Presenting discoveries about how to draw figures on the graphing calculator

Outline of the Day

In Class

1. Discuss *Homework 1: Starting Sticky Gum*
   • Let students try to handle one another's questions
2. Continue work on *Picture This!*
   • Students continue to work with manuals and calculators and to share their discoveries
   • Make sure students know how to clear the calculator screen and draw circles and line segments
3. Introduce writing a program to make a drawing

At Home

Homework 2: Programming Without a Calculator

Special Materials Needed

• A calculator manual for each pair of students

Discuss With Your Colleagues

Calculators and Equity

This and other IMP units are written so that graphing calculators are not needed for homework assignments, because students might not have access to graphing calculators at home. But in this unit, having a graphing calculator at home is particularly useful because of *POW 5: An Animated POW*. So what happens if some of your students do have such home access? Does that give them an unfair advantage? What can you and your school do to create equity among all students in this respect?

Day 2

1. Discussion of *Homework 1: Starting Sticky Gum*

Begin with a presentation on Question 1 of the POW, perhaps having several students contribute. One goal of this discussion is to have students see a model of a general formula for such a problem. Perhaps the simplest approach to this question is to examine what the maximum amount is that a parent could spend and *not* get the desired result. For instance, suppose there are n children and c colors. The worst that could happen, without getting n gum balls of the same color, is for the parent to get $n - 1$ gum balls of each color. This would cost $(n - 1)c$ cents, so the maximum that might be needed is $(n - 1)c + 1$ cents.

- Question 2

    Go over Question 2 to be sure students see how this problem differs from the situation in Question 1. Here, the worst-case scenario is for Ms. Hernandez to get all eight yellow gum balls before getting anything else. Thus, the most she might have to spend is nine cents.

- Question 3

    Ask if anyone has questions about what is expected in this POW. Let other students answer the questions, and clarify as necessary.

    Emphasize that students' write-ups for the POW should focus on general conclusions for problems like Question 2. (The write-ups should also include Questions 1 and 2, even though students have already written these up for homework, and they should include Question 3 of the POW. Students may want to revise their work from Questions 1 and 2 based on today's discussion.)

2. Continued Work on *Picture This!*

Have students continue their investigation of how to create drawings on the graphing calculator and their presentations of their discoveries.

As indicated yesterday, you need to make sure that the presentations include these items.

- How to clear the calculator screen
- How to draw on the screen a line segment that connects two points that are given in terms of their coordinates
- How to draw on the screen a circle that is given in terms of its radius and the coordinates of its center

Except for these items, no specific features need to be discussed. (Students will need the line command in tonight's homework.)

Day 2

Bring out that because the "line" and "circle" commands involve the calculator's coordinate system, students need to adjust their viewing rectangles to see the results of these commands. Also, you may want to discuss the fact that unless the scales on the x- and y-axes are the same, circles described in terms of coordinates will come out looking like ellipses. (*Note:* Having the viewing window go from, say, -10 to 10 for both x and y will *not* make for equal scales unless the actual screen is square, which is generally not the case.)

3. Writing a Program

As needed, briefly review the general idea of writing a program for the graphing calculator. (Students wrote programs as part of the Year 1 unit *The Game of Pig*.)

Then illustrate (or have a volunteer illustrate) how to incorporate into a program some of the basic drawing features that students have discussed. For instance, you might show students how to write a program to draw two line segments. Be sure to explain how to run the program, and bring out that students may need to adjust the viewing rectangle to see the results of the program.

If time allows, have students write short programs of their own to make some drawing.

• "Code" and "plain language"

Bring out that every programming language uses very specific syntax and commands. This formal programming language is often called **programming code** (or simply *code*). Emphasize that the code for a particular task is likely to vary from one calculator model to another. (You might mention that there may also be more than one way to write the code for a particular calculator.)

Tell students that programmers often begin with descriptions in ordinary language of what they want to do and then turn those descriptions into programming code. We will refer to such descriptions as **plain-language programs.**

Illustrate the distinction between coded programs and plain-language programs with one or two examples. For instance, you can compare the statement "draw a line segment from $(4, 2)$ to $(7, 5)$" with the programming command that students need to use on their specific calculator to accomplish this task.

Tell students that some of their assignments in this unit will involve plain-language programs, including these types of tasks.

- Describing what a particular plain-language program will do when run
- "Translating" a plain-language program into code for their specific calculator model
- Creating a plain-language program to accomplish a specific task

In tonight's homework, they will perform tasks of the first two types, as well as enhance their code program to make it more interesting. Take a few minutes to go over the homework to be sure students understand what is expected in each part of the assignment.

Comment: When students do their animation projects (in *POW 5: An Animated POW*), they will also be asked to turn in an *outline* of their program. We use this term to mean a description that is more of an overview than the plain-language program. On Day 9, students will create an outline for a "generic" animation program. At that time, you can discuss the distinction between a program and an outline.

Homework 2: Programming Without a Calculator

> In tonight's homework, students will get some practice using the instruction to draw line segments and will translate a plain-language program into code. They will then add their own touches to the program.

Day 2

Programming Without a Calculator

Every programming language uses very specific syntax and commands. This formal programming language is often called *programming code,* and the code for a particular task is likely to vary from one calculator model to another.

Programmers often begin with "plain-language" descriptions of what they want to do. Then they turn those descriptions into programming code.

Over the course of this unit, you will have several assignments asking you to write or interpret plain-language programs. Each of these programs will begin with a title line, and each command will begin on a new line. You will also sometimes be asked to turn a plain-language program into programming code, as in Question 2 of this assignment.

1. Read through the steps of the plain-language program shown on the next page, and then draw on graph paper what should appear on the calculator screen when the program is run. (Assume that the calculator has an appropriate viewing rectangle.)

Continued on next page

Interactive Mathematics Program   *As the Cube Turns* 119

Program: LINES

Clear the screen

Draw a line segment connecting (−4, −2) to (2, −2)

Draw a line segment connecting (−4, 2) to (−1, 3)

Draw a line segment connecting (−1, 3) to (1, 5)

Draw a line segment connecting (2, −2) to (4, 0)

Draw a line segment connecting (1, 5) to (4, 4)

Draw a line segment connecting (1, 5) to (−2, 4)

Draw a line segment connecting (−4, 2) to (−2, 4)

Draw a line segment connecting (4, 4) to (4, 0)

Draw a line segment connecting (−1, 3) to (2, 2)

Draw a line segment connecting (−4, 2) to (−4, −2)

Draw a line segment connecting (2, 2) to (4, 4)

Draw a line segment connecting (2, −2) to (2, 2)

Note: If you don't get something that looks like a real picture, you've probably made a mistake somewhere and should check your work.

2. Turn the plain-language program of Question 1 into programming code. In other words, write the program lines you would enter into your calculator to get the drawing you got for Question 1.

3. Add some more commands to the program you wrote in Question 2 in order to improve the rather dull picture. Then draw what should appear on the calculator screen when you run your improved program.

Day 3

Programming Loops

This page in the student book introduces Days 3 through 7.

Rebecca Yaeger writes a report about her work so far in determining how to rotate a cube.

You may think of a loop as something you make with a shoelace or a piece of string, but it is also a handy device in writing programs for computers or calculators. Loops allow you to repeat the same set of instructions as many times as you like.

Over the next several days, you will use loops to simplify the work of writing a program and to create the illusion of motion.

Interactive Mathematics Program   As the Cube Turns   121

DAY 3

Around and Around We Go!

Students learn to use loops in programs.

Mathematical Topics

• Working with programming loops

Outline of the Day

In Class

1. Discuss *Homework 2: Programming Without a Calculator*
   • Have students enter their code into graphing calculators
2. Introduce loops
   • Introduce loops and their syntax as a shortcut for repetitive programs
   • Have students act out what happens inside the calculator as it executes a loop
   • Have students enter a loop program into their calculators
3. Using the For/End loop variable
   • Illustrate how to use the loop variable as more than a counter

At Home

Homework 3: *Learning the Loops*

Special Materials Needed

• (Optional) A box and slips of paper for students to "act out" loops
• A transparency of the screen diagram used for introducing For/End loops (see Appendix C)

1. Discussion of *Homework 2: Programming Without a Calculator*

• Question 1

Have a volunteer prepare a transparency to show the diagram for Question 1. Meanwhile, have some students enter the code they wrote for Question 2 into their calculators and the rest enter the code they wrote for Question 3.

Interactive Mathematics Program   As the Cube Turns   23

Day 3

Here is a sample of what the diagram for Question 1 might look like. (This diagram is different in some respects from what will appear on the calculator screen. If there are substantive differences of opinion about this diagram, you should discuss them, but don't get sidetracked by technical details.)

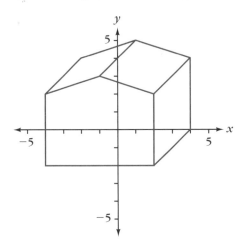

- Question 2

Then have students who entered their code for Question 2 run their programs, and compare their results to their diagrams for Question 1. They will probably see some differences, including these.

- Some of the line segments on the calculator screen may appear rather jagged. (If needed, you can explain that this is due to the use of individual pixels.)

- The calculator screen will probably not label the axes or show numerical values along the axes (although it may show tick marks on the axes).

- The drawings may not be centered on the calculator screen, due to the choice of viewing window.

- The diagram may appear "stretched" either vertically or horizontally, because the unit distances along the two axes are not equal.

You should discuss these differences with the class. Throughout this teacher guide, we will represent such calculator results using diagrams that look more like ordinary graph paper diagrams than like calculator screens. We suggest that you have students use similar graph-paper style representations (including axis labels and scales) when they are describing the output of a program. But make it clear to students that if they are asked to write a calculator program to produce a diagram like the one just shown, they are not expected to get the calculator to label the axes, mark the scales, or create "smooth" line segments. However, they will have to choose appropriate viewing windows in order to make their results visible on the screen.

Day 3

If students see substantial differences between the predicted and the actual results other than those just discussed, it is likely they have made errors in their code, and they should locate these errors.

Have one or two students who got a result like what they expected for Question 2 present their code and show the results on the overhead. You may have some variations in the code that students use, so allow time for discussion.

• Question 3

If time permits, have some volunteers present their enhanced programs from Question 3, both running the programs and explaining their code.

2. The Basics of Loops

Comment: Students may recall their experience writing and using loops in a simulation for the newspaper carrier problem in the Year 1 unit *The Game of Pig* (see Day 21 of that unit). Some students may have other programming experience as well. Nevertheless, we have taken a cautious approach, presenting these ideas as if they are being introduced for the first time. You will need to make adjustments to fit the needs of your class.

The next stage in the unit is to look at the combination of instructions needed to write a loop. We suggest that you begin with a simple example to motivate the need for loops.

"Can you create a program that will show this display on the screen?"

```
HI
BYE
HI
BYE
HI
BYE
```

Ask students to work in their groups to write a program on the calculator that will generate a screen display like the one shown at the left.

Students may have discussed (on Day 1 or 2) the instructions for getting the calculator to display text on its screen. If not, then you will need to introduce the necessary code.

Students will probably create a program that contains a separate instruction for each line of the desired display. In plain-language form, the program might look like this:

> Display the word "HI" on the screen
>
> Display the word "BYE" on the screen
>
> Display the word "HI" on the screen
>
> Display the word "BYE" on the screen
>
> Display the word "HI" on the screen
>
> Display the word "BYE" on the screen

Day 3

Comment: This plain-language program ignores details such as the exact position of each display. You may need to include those details in the discussion of the specific programming code. Often, each new display instruction will place the desired text at the beginning of a new line, which is the goal here.

- *The plain-language "For/End" loop*

"How might you avoid the repetition in the program?"

Bring out the fact that the same pair of instructions is being repeated three times, and ask if students know a way to avoid this repetition. Tell them that most programming languages have a mechanism (and perhaps several different mechanisms) for doing the same set of instructions more than once.

If you can use students' ideas to develop this loop, fine. Otherwise, have a student give you a letter to use as a variable. Then show the class this plain-language program:

```
For A from 1 to 3
Display the word "HI" on the screen
Display the word "BYE" on the screen
End the A loop
```

Tell students that this plain-language program uses the instructions "For" and "End" as a generic way to create a programming device called a **loop.** Also introduce the related terminology. In this example, the letter A is called the **loop variable** and the numbers 1 and 3 are called, respectively, the **initial value** and the **final value** for the loop variable. The two lines between the "For" instruction and the "End" instruction form the **body** of the loop.

Tell students that to show the overall structure of the plain-language program more clearly, we sometimes indent the body of the loop, with a bullet for each step of the body. (This convention is mentioned in tonight's assignment, and we will generally follow it.) For instance, the four-line program just shown might be written like this:

```
For A from 1 to 3
    • Display the word "HI" on the screen
    • Display the word "BYE" on the screen
End the A loop
```

- *Acting out the program*

"How do you think this program works?"

Ask students how they think this program works. Their answers may be somewhat speculative, but some students likely will be able to describe the general idea.

After some verbal descriptions, have the class act out the program, perhaps using a box drawn on the chalkboard (or an actual box) as the "memory cell"

Day 3

for the variable A. (Students should be familiar with the idea of a memory cell for a variable from earlier work with programming.) You should also set up something to serve as the "calculator screen."

Although calculators may vary slightly in how they actually carry out a loop like this, we suggest that you have students act it out using this scenario:

- They start with the "For" instruction, and interpret that by putting the number 1 in the box labeled A.

- They carry out the body of the loop, which in this case consists of the two display instructions. So they should print "HI" and then "BYE" on the "screen."

- For the "End" instruction, they increase the value of the variable A (that is, the number in box A) to the next value (which at this stage means going from 1 to 2) and then go back to the "For" instruction.

- When they return to the "For" instruction, they check to make sure that the loop variable does not exceed 3. It does not (because A is now 2), so they proceed.

- They carry out the body of the loop, printing "HI" and then "BYE" again.

- This time, when they reach the "End" instruction, they increase the value of A from 2 to 3, and again go back to the "For" instruction.

- When they return to the "For" instruction, they again check to make sure that the loop variable does not exceed 3. It does not (because A is now 3), so they proceed.

- They carry out the body of the loop, printing "HI" and then "BYE" again (for the third time).

- This time, when they reach the "End" instruction, they increase the value of A from 3 to 4, and again go back to the "For" instruction.

- When they return to the "For" instruction, they check to make sure that the loop variable does not exceed 3. This time it does (because A is now 4), so the loop is completed.

Bring out that if there were any instructions after the "End" instruction, students would proceed by going to the first such instruction. In this example, there are no such instructions, so the program is over.

> We suggest that you leave questions such as "What's the value of A after you complete the loop? Is it 3 or 4?" and "What happens if you use an instruction like 'For A from 1 to 2.5'?" for interested students to investigate on their own. In *Homework 16: Taking Steps,* students will look at some other issues about loop variables.

Day 3

- *Writing the program in code*

Go over how students can write this simple loop in code for their calculators. There may be more than one option, but many calculators have a loop format that uses the specific instruction words "For" and "End." We recommend that you have students use this type of loop if possible.

Then have students enter and run the program on their graphing calculators to verify that it works.

> *Note:* In some calculators, certain letters serve special purposes, such as using x and y for defining functions. You need to warn students never to use such letters as variables in their programs.
>
> Also, it's easy to accidentally write a program that sets up an infinite loop. There should be a key on your calculator to interrupt a program while it is running. You may need to introduce that key to your students.

3. Using the For/End Variable

"What role does the variable A play in this program?"

Ask the students what role the variable A plays in this program. They should see that it basically keeps track of how many times the pair of instructions to display "HI" and "BYE" is carried out. In such an example, the loop variable is sometimes called a **counter.**

The next step in understanding the use of the For/End loop is to see how the loop variable can be used in a way beyond simple counting. Show the class the accompanying diagram. (A large version of this diagram is included in Appendix C for your use in making an overhead transparency.)

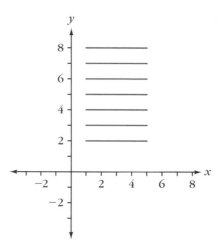

"What program can you write to produce these line segments on the screen?"

Have students work in groups to write a program to produce these line segments on the calculator screen. Be sure they realize that they do not need to get the calculator to label their axes or show numerical scales (but they do need to set the calculator to show an appropriate viewing window).

Day 3

Some students may recognize that they can use a loop to do this, but others may not. For instance, many students probably will come up with something analogous to this plain-language program:

Draw a line segment from (1, 2) to (5, 2)
Draw a line segment from (1, 3) to (5, 3)
Draw a line segment from (1, 4) to (5, 4)
Draw a line segment from (1, 5) to (5, 5)
Draw a line segment from (1, 6) to (5, 6)
Draw a line segment from (1, 7) to (5, 7)
Draw a line segment from (1, 8) to (5, 8)

"How might you use a loop to shorten the program?"

If no one suggests the use of a loop, ask students how a loop might be used to shorten the program. As a hint, point out that these instructions are all quite similar, or ask what is changing in the program. Someone should be able to come up with the idea of a program like this:

For B from 2 to 8
 • Draw a line segment from (1, B) to (5, B)
End the B loop

You may want to have students act out how this works, step by step. At least go over the fact that the initial value is 2, so the calculator begins by putting the number 2 in a memory cell labeled B.

"What is the y-coordinate for the first line segment that needs to be drawn?"

One aspect of this program that may cause some difficulty is the fact that the loop variable is starting at a value other than 1. As a hint, you can bring out that the *y*-coordinate is changing from one line of the original program to the next, and ask what the *y*-coordinate is for the first line segment that needs to be drawn.

Comment: Students who want to use 1 for the initial value can do so by having the body line of the loop be

 • Draw a line segment from (1, B+1) to (5, B+1)

In this case, the "For" instruction would read

For B from 1 to 7

If someone suggests this way out, that is fine. But make sure students are aware of the option of changing the initial value of the variable.

Interactive Mathematics Program   As the Cube Turns 29

Day 3

• *Entering and running the program*

After this discussion, have students enter and run the new program and check that it works properly. As with all drawing programs based on coordinates, students may need to adjust the viewing rectangle to get the desired picture. This is a good occasion to look at how to set the viewing rectangle within the program itself. (This discussion can be delayed and included in connection with the "setup" program discussed on Day 6.)

Homework 3: Learning the Loops

This homework will give students practice using loops in programs. Tomorrow's discussion will help you to assess your students' background and comfort with regard to programming.

Carmen Rubio has written out her program instructions and is now ready to enter them in her graphing calculator.

Learning the Loops

The For/End combination of instructions can be used to do lots of interesting things. Your job is to figure out how it all works, both in the plain-language version and in the version using programming code for your calculator.

Note: To show the overall structure of the program more clearly, it's helpful to indent the body of the loop. The plain-language programs in this and later assignments use that format, and start each step of the body of the loop with a bullet (the symbol •).

1. Describe what will happen when you run a calculator program based on this plain-language program.

Program: LOOP1

For T from 1 to 5
   • Display "HELLO" on the screen
End the T loop

2. Describe what will happen when you run a calculator program based on this plain-language program.

Program: FRUITLOOP

For A from 1 to 5
   • Display "LEMON" on the screen
   • For B from 1 to 3
      • Display "LIME" on the screen
   • End the B loop
End the A loop

Continued on next page

3. Sketch what the calculator should show when you run a program based on this plain-language description.

   Program: LOOP2

   For G from 3 to 10
   • Draw a line segment from (G, G) to (G+3, G+1)
   End the G loop

4. Write a short program (in programming code, using a For/End loop) that you think will draw something interesting. Describe in words what you think it will draw.

DAY 4

Animation

Students make their first animation program.

Mathematical Topics

- Continuing to work with programming loops
- Creating animation on the graphing calculator

Outline of the Day

In Class

1. Discuss *Homework 3: Learning the Loops*
   - Act out Question 2
   - Introduce the term **nested loop**
   - Let students share their original programs
2. Discuss where the class is in the unit
   - Students should see that they have completed the first item in the unit outline
3. Introduce flip book animation
   - (Optional) Show students a sample flip book
4. *An Animated Shape*
   - Students learn to create a sense of movement on the calculator screen
   - The activity will be discussed on Day 5

At Home

Homework 4: A Flip Book

Special Materials Needed

- Six to eight index cards per student for homework
- (Optional) A sample flip book

1. Discussion of *Homework 3: Learning the Loops*

- Question 1

   Question 1 should be straightforward. Bring out that the loop variable is acting simply as a counter.

Day 4

• *Question 2*

Have students act out this program. They will need two boxes—one for each of the two variables. The program should create an output sequence like this:

LEMON
LIME
LIME
LIME
LEMON
LIME
LIME
LIME
LEMON
LIME
LIME
LIME
LEMON
LIME
LIME
LIME
LEMON
LIME
LIME
LIME

Note: If the program is translated into code and run on a calculator, it may be impossible to confirm that there are the right number of LEMONs and LIMEs in the display, because the display will go by too fast to count the LEMONs and LIMEs as they are printed on the screen, and only the last few lines of the display will be visible on the screen when the program is over.

Introduce the term **nested loop,** which means a loop within a loop, for the programming structure involved in Question 2.

"How would you write the code for this program?"

Ask students how they would write the code for this program. Bring out that in our plain-language programs, the "End" instruction states what loop it refers to, but in most actual programming code, it does not. Clarify that the general convention is that an "End" instruction sends the program back to the most recent "For" instruction.

- Question 3

    For Question 3, ask for volunteers to describe what they think will happen. (It's important that students feel comfortable about making these predictions. You can assure them that even experienced programmers often make mistakes in interpreting instructions.) You may want to have students act it out as they have with other examples.

    Then have students turn this plain-language program into code and try out their programs to confirm their predictions. Of course, the result on the screen will depend on the viewing rectangle, and the line segments will probably appear rather jagged due to the use of individual pixels. However, the result should look something like this:

    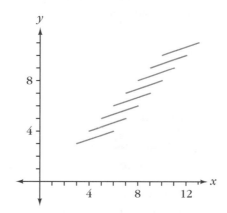

- Question 4

    For Question 4, we suggest that you have students enter their programs into their calculators and perhaps share their work within their groups. You might have some students show their programs to the class using the calculator overhead projector.

2. Where Are We in the Unit?

Remind students that their goal is to create an animated picture of a cube turning. Summarize that they now know something about drawing on the calculator screen and that they are learning about the programming language of the calculator so that they can write appropriate instructions.

Day 4

Refer to the plan posted on Day 1. Students can now check off item 1: "Draw a picture on the graphing calculator." Then review the remaining items, which you might organize into three categories.

- Creating the illusion of motion on the calculator screen
- Changing the position of an object located in terms of a coordinate system
- Creating a two-dimensional drawing of a three-dimensional object

Over the next few days, we will look at how to create the illusion of motion. We will then consider how to change the position of two-dimensional objects (Days 8–19), how to create a two-dimensional drawing of a three-dimensional object (Days 20–30), and how to change the position of three-dimensional objects (Days 31–33).

This is a good time to review the terms **2-space** and **3-space** to refer to the sets of all points in the two-dimensional and three-dimensional coordinate systems.

3. Flip Books: The Illusion of Motion

"What is animation? How does it work? How could you create it?"

Ask if anyone can explain the basic idea of animation. Bring out these two key elements.

- Animation involves a series of drawings, each differing slightly from the previous one.
- The illusion of motion is created by showing each picture very briefly and then replacing it by the next.

Tell students that one simple device for showing animation is a *flip book*. If possible, illustrate with an example (perhaps one made by a former student). If you have no example available, describe how a flip book can be made with a set of index cards, illustrating how to hold and flip through the cards so that each shows for only a fraction of a second. Tell students that they will create their own flip books in tonight's homework assignment, using a set of index cards, and will apply the same principle in today's activity, *An Animated Shape*.

4. *An Animated Shape*

In *An Animated Shape,* students begin work on creating a sense of movement on the calculator screen. This is a good activity for students to work on in pairs.

Stress that students should not get so carried away with artistry on their initial picture (Question 1) that they don't have time to complete the remaining parts of the activity. They will have additional time tomorrow to finish work on this activity and discuss it.

Day 4

Students will probably realize that they need to erase the screen before drawing each new picture. But they may not realize that they need to delay before doing so in order to allow time for the picture to register on the brain. The mechanics of introducing a "delay" into a program should come up in the discussion of this activity, and students need not develop it now. But if you want to plant the seed of that idea, you can ask pairs to think about why they see only one picture (the last one) when they run their programs.

Note: We use the term *delay* rather than *pause* because in some calculators a "pause" command simply halts the program until the user tells the calculator to resume. That process will not create an effective animation program.

Students who recognize the need for a "delay" may become frustrated if they are unable to get the calculator to do what they want about this. If they can't come up with a solution to this problem now, assure them that this will be included as part of the discussion of the activity.

Homework 4: A Flip Book

In this assignment, students create flip books to further understand the idea of animation.

Give students index cards to use in making their flip books.

Interactive Mathematics Program   As the Cube Turns 37

Day 4

CLASSWORK

An Animated Shape

Your task in this activity is to write a program to create and animate a shape on your calculator. As you progress, write down any questions you have about writing programs for animation. Keep in mind that you are trying to create the illusion of motion.

1. First, draw your shape on graph paper. Make it very simple! Your picture should have no more than five line segments or pieces.

2. Next, write out the program code to have the calculator draw your shape. (You may want to begin with a plain-language program.)

3. **a.** Draw the same shape on graph paper in a new location, near the first.

   b. Write additional program instructions to make it appear that your shape has moved to its new location.

   c. Repeat steps 3a and 3b several times, changing the position of your shape slightly each time.

4. Enter your program into the calculator and run it. Locate and correct errors if the program doesn't run as you would like.

A Flip Book

A *flip book* is a device consisting of a series of pictures, each slightly different from the previous one. The pictures are drawn on index cards (or something similar) so that you can flip through them to create the illusion of movement like animation.

1. Create a simple flip book of your own. (You don't have to be a great artist to complete this assignment. A moving rectangle or a rolling ball is okay. On the other hand, if you like drawing, let yourself have fun with this.)

2. Explain how you think this assignment is related to the unit.

DAY 5

Students learn how to let their calculators take a break.

Busy Work

Mathematical Topics

- Using a flip book to demonstrate animation
- Creating animation on the graphing calculator
- Inserting a delay loop in a program

Outline of the Day

In Class

1. Discuss *Homework 4: A Flip Book*
   - Have students share their work within groups
   - Tell students to begin thinking about their project
2. Complete and discuss *An Animated Shape* (from Day 4)
   - Bring out that animation involves drawing, delaying, erasing, and redrawing
3. Discuss how to create a "delay loop"

At Home

Homework 5: Movin' On

1. Discussion of *Homework 4: A Flip Book*

Rather than discuss *Homework 4: A Flip Book*, you might simply give students a few minutes to share their flip books with one another within groups.

Interactive Mathematics Program

Day 5

- *A project reminder*

    While students are sharing flip books, remind them that this unit includes a final project in which they will work with a partner to create an animation program on the graphing calculator. Point out that although they may not know enough yet to begin writing the program, they can begin thinking about a subject for their project and considering who to work with.

2. Completion and Discussion of *An Animated Shape*

Allow the class additional time to continue work on *An Animated Shape*. When most students seem ready for a discussion, bring the class together.

"What worked and what didn't work?"

You can begin by asking students to describe what their results were, perhaps having one or two pairs demonstrate their programs on the calculator overhead projector. Also have them share any mistakes they made along the way and discuss any difficulties they encountered.

Note: You should leave time today for discussion of how to include a "delay" in programs (see next section). If you want to have more students demonstrate their programs, you can do this tomorrow.

"How did you create the illusion of motion?"

As part of the discussion, ask what steps were involved in creating the illusion of motion. Three key issues need to be identified.

- They need to erase their first picture before drawing the second one.
- They need to know how to show the same picture in a different position.
- They need to delay before erasing each successive drawing. Otherwise, all drawings but the last will blur by very quickly, and they will really see only the last drawing.

Most likely, students will have figured out that they need to include an instruction in their program to erase the previous drawing. But they may have some trouble with the details of the other two issues, especially the question of how to make the calculator delay before erasing.

The details about how to move an object are the essence of items 3 and 5 in the unit outline, and students will begin work on these details on Day 8.

3. The Delay

"How can we get the calculator to delay?"

Ask for suggestions about how to create a "delay" in a program.

One solution (which we will use in this teacher guide) is to have the calculator run through a loop with no body instructions. As a hint for students to suggest this idea, you can propose that they give the calculator

Day 5

something to do to keep it busy before it clears the screen. One such task would be to have it count to some number such as 200. Here is the basic structure of such a program:

> For T from 1 to 200
>
> End the T loop

We will refer to this as a **delay loop.** Bring out that students can control the duration of the delay by adjusting the final value of the loop variable.

Homework 5: Movin' On

> You can use this homework to assess and grade students on their understanding of loops.

Sharon Wang and Jeff Tung are pleased about having successfully programmed their graphing calculator to create the illusion of motion.

Movin' On

It's often helpful to have a "setup" program at the beginning of a calculator program. This setup program might adjust the viewing window, clear the screen, and so on.

From now on, we will include a line that simply says "Setup program" in all plain-language programs that involve screen graphics. When you translate such programs into code, you will need to give details for the setup program.

1. **a.** Use drawings on graph paper to describe the result of this plain-language calculator program.

   Program: SEGMENTS

   Setup program

   For S from 1 to 5
   - Clear the screen
   - Let T be 3 more than S
   - Let U be 6 more than S
   - Draw a line segment from (S, S) to (T, T)
   - Draw a line segment from (T, T) to (U, S)
   - Draw a line segment from (U, S) to (S, S)
   - Delay

   End the S loop

   b. Create programming code for the plain-language program in Question 1a. Include an appropriate setup program. (You can write the setup program using plain-language instructions, if necessary.)

2. Make a general list of components that you would include in a setup program. (This may include items that were not needed in Question 1b.)

DAY 6

Set It Up!

Students develop ideas for a "setup" program.

Mathematical Topics

- Analyzing what a specific program will do
- Doing animation in a program using a loop
- Developing a setup program

Outline of the Day

In Class

1. Select presenters for tomorrow's discussion of *POW 3: "A Sticky Gum Problem" Revisited*
2. Discuss *Homework 5: Movin' On*
   - Have students enter their code to check if it works
   - Have students share ideas on what should be included in a setup program
3. Continue with demonstrations of programs from *An Animated Shape* (begun on Day 4)

At Home

Homework 6: Some Back and Forth

1. POW Presentation Preparation

Presentations of *POW 3: "A Sticky Gum Problem" Revisited* are scheduled for tomorrow. Choose three students to make POW presentations, and give them overhead transparencies and pens to take home to use for preparing presentations.

These discussions should focus on situations like Question 2. If that specific problem was dealt with fully on Day 2, then the POW presentations should begin with Question 3 and then turn to generalizations.

Day 6

2. Discussion of *Homework 5: Movin' On*

- *Question 1*

    Let a volunteer present Question 1a, which should involve a sequence of diagrams like this:

    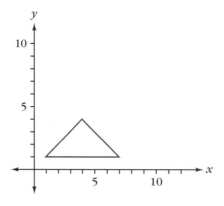

    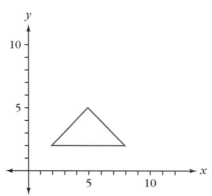

    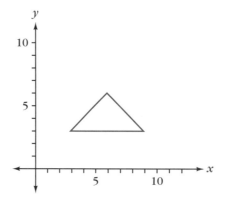

Day 6

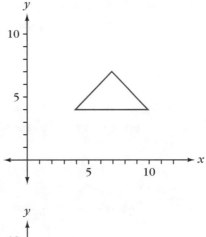

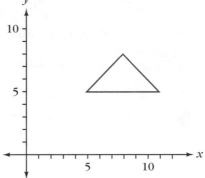

Give students some time to enter the code they created for Question 1b to see if it gives the desired result. Then have one or two students present successful code (or discuss this as a whole class if no one was able to create the appropriate code).

• Question 2

"Why might a setup program be needed or desirable?"

You might begin the discussion of Question 2 by asking students to share ideas on why a setup program might be needed or desirable.

Students have probably seen that when they run programs involving a coordinate display, they often need to adjust the viewing window. Some students may also have run into the dilemmas of having functions graphed over their work or of having to clear an old graphic from the screen.

Bring out that a setup program can be used to eliminate these stumbling blocks. Here are two ways students can incorporate such a program into their work.

- They can write a separate program called "Setup" and have "Run the setup program" be the first instruction of each drawing program they write.

- They can copy the individual instructions of the setup program into the beginning of each drawing program they write.

Interactive Mathematics Program   As the Cube Turns 47

Day 6

"What instructions should be included in a setup program?"

Have the class suggest instructions that should be included in this setup program. Here are some possibilities.

- Clear the screen.
- Make the function graphing feature inactive.
- Set the viewing rectangle (the parameters in this part of the setup program will need to be adjusted to suit the individual program).

> *Note:* If the "drawing" portion of the program involves a loop that starts with an instruction to clear the screen, then students need not include that instruction in their setup program. (On the other hand, it certainly won't hurt to do so. If students want to have a standard setup program, they might include a "clear screen" instruction even if it's sometimes redundant.)

Have students develop code for each of these components, and test it on a program. They should also keep a written record of these ideas in their "programming notebook."

Students will no doubt encounter other items to include in the setup program as they progress through the unit. For instance, when they start working with angles, they may want to have an instruction that sets the "mode" for angles to degree measurement.

3. More Presentations from *An Animated Shape*

If you have additional time, you can let students who did not present their work from *An Animated Shape* do so today.

Homework 6: Some Back and Forth

> This assignment continues the ideas of *Homework 5: Movin' On,* but now students need to include a description of their setup program as well.

Some Back and Forth

1. The sequence of graphs here shows a line segment going back and forth between two positions. Use a loop to write a plain-language program to produce an animation showing the line segment in these positions, one position after the other. (Your program does not need to draw the coordinate axes or scales. They are shown here merely to indicate the changing position of the line segment.)

First position of segment:

Second position of segment:

Third position of segment:

Fourth position of segment:

Continued on next page

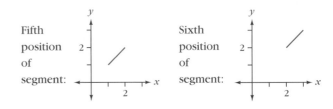

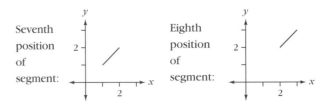

2. Create programming code for the plain-language program you wrote for Question 1, including an appropriate setup program.

DAY 7

POW 3 Presentations and a Moving Arrow

Students present POW 3 and continue creating animation in programs.

Mathematical Topics

• Doing animation in a program using a loop

Outline of the Day

In Class

1. Presentations of *POW 3: "A Sticky Gum Problem" Revisited*
   • If students disagree with answers, have them look for flaws in the proofs
   • Generate a list of class generalizations and conjectures
2. Discuss *Homework 6: Some Back and Forth*
   • Have students try out their setup programs
3. Arrow
   • Students use a delay loop to create a moving arrow
   • No whole-class discussion is needed for this activity

At Home

Homework 7: Sunrise

POW 4: A Wider Windshield Wiper, Please (due Day 20)

1. Presentations of *POW 3: "A Sticky Gum Problem" Revisited*

Ask the three students to make their presentations, building on what was accomplished in the Day 2 discussion of *Homework 1: Starting Sticky Gum*.

"If you disagree with an answer, what flaw can you find in the reasoning?"

Because the emphasis in this POW is on proof, focus students' attention on finding any errors in the presenters' proofs. List generalizations about these problems on the chalkboard or on chart paper. You may have two lists—proved generalizations and conjectured generalizations—with the second group being those for which no one has given a convincing proof.

Spend as much time on this problem as student interest dictates.

Day 7

2. Discussion of *Homework 6: Some Back and Forth*

Begin with a discussion of the plain-language program, and have one or two students present their ideas. Here is one option:

> Setup program
>
> For A from 1 to 4
> - Clear the screen
> - Draw a line segment from (1, 1) to (2, 2)
> - Delay
> - Clear the screen
> - Draw a line segment from (2, 2) to (3, 3)
> - Delay
>
> End the A loop

You can then have students enter their programming code into the calculators to see if it creates the desired sequence of line segments. This is a good opportunity for students to try out their ideas about setup programs.

3. *Arrow*

Have students begin work on the activity *Arrow*. This activity probably needs no discussion unless students want to share embellished programs using the calculator overhead projector.

Homework 7: Sunrise

> This assignment gives students a chance to apply the new ideas about animation, as well as to use their calculator's capacity for drawing circles.

Take a few minutes to remind students of the programming code for drawing circles on their calculators. The code will probably use three parameters: two for the coordinates of the center and one for the radius.

POW 4: A Wider Windshield Wiper, Please

Make sure students realize that the report they write for this POW is like a business analysis, not the standard POW write-up. You may want to set some guidelines about the length of these reports. You might also suggest to students that they actually examine the windshield wipers on some cars to see how they function.

Day 7

Arrow

This plain-language program describes the animation of a flying arrow. Create programming code for this plain-language program, enter the code into your calculator, run the program, and then find ways to improve it.

Program: ARROW

Setup Program

For A from 1 to 8
- Clear the screen
- Make B 3 less than A
- Make C 1 less than A
- Draw a line from (B, B) to (A, A)
- Draw a line from (C, A) to (A, A)
- Draw a line from (A, C) to (A, A)
- Delay

End the A loop

Sunrise

In this assignment, you will create a program that displays a rising sun by showing a circle "moving" up and across the screen.

There are three parts to this assignment.

- Make a sequence of drawings on graph paper to show what you want to appear on the screen. You should show at least four different positions for the circle.

- Write a plain-language program describing how to create your sequence of drawings.

- Write programming code for your plain-language program. Be sure to include settings for the viewing window so that your circles will be visible.

A Wider Windshield Wiper, Please

Imagine! Straight out of high school and you land a job at Better Design Ideas, Inc. Your boss is just starting to explain the company's project to design a car windshield wiper that will clean more area than the standard wiper.

The standard wiper consists of a 12-inch blade attached rigidly to a 12-inch arm. The middle of the blade is at the end of the arm, and the arm rotates back and forth, making a 45° angle with the horizontal at each end of its motion. The path swept out by the blade of the standard wiper is shown by the shaded area in the first diagram on the next page.

While your boss is talking, two men in white coats rush in with another model. "It's better!" exclaims one. "It's worse," shouts the other.

Continued on next page

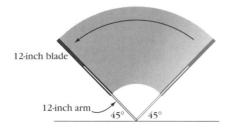

The new model also has a 12-inch blade attached at its midpoint to a 12-inch arm. But in this case, the blade pivots so that it's always vertical. The arm rotates 90° as in the standard model. The next diagram shows the path swept out by the blade of the new wiper.

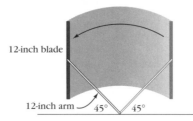

The two men keep shouting, but your boss merely rolls her eyes, hands you a white coat, and tells you to find out who's right. Just as you are racing out the door, she calls after you, "In any case, make a better one—one that will clean more area."

You know you are limited to 12-inch blades, which are an industry standard, and that any rotating arm can rotate only 90°. You also know your job is at stake, so go to it!

As your write-up for this POW, write a report for the boss. Be sure to state any further assumptions that you make.

Day 8

DAYS 8-10
Translation in Two Dimensions

This page in the student book introduces Days 8 through 10.

Rachel Silverman has successfully programmed her graphing calculator to make a straight line appear animated.

You have now learned the principles of how to create the appearance of motion and learned how to create programming code to carry out some basic animation. Part of this work involved translating plain-language programs into programming code.

Now you're going to work with another meaning for the word *translation*—a geometric meaning—that will help you carry out a specific animation task. You'll also see how to use matrices to assist with this task.

DAY 8

Move That Line!

Students learn how to translate graphics.

Mathematical Topics

• Introducing geometric translations

Outline of the Day

In Class

1. Discuss *Homework 7: Sunrise*
   • Review the use of setup programs
   • Bring out that students have completed the second item in the unit outline

2. Introduce geometric transformations
   • Introduce the terms **transformation** and **translation**
   • Develop the representation of a translation by a **translation vector**

3. *Move That Line!*
   • Students use delays and loops to move a segment repeatedly by the same translation
   • The activity will be completed and discussed on Day 9

At Home

Homework 8: Double Dotting

1. Discussion of *Homework 7: Sunrise*

Have students enter their programs into the calculators and see how they work. While students are working, have one or two students put their plain-language programs on a transparency to share. Also have a couple of students use the calculator overhead projector to display programs to the entire class.

If it hasn't yet come up, bring out that unless the scales on the x- and y-axes are the same, circles will come out looking like ellipses.

• *Another look at setup programs*

The issue of scales on the axes provides another context in which to discuss the use of a setup program (see Day 6). Here you can focus specifically on

Interactive Mathematics Program   As the Cube Turns   **59**

Day 8

the issue of setting the viewing window, which will vary from program to program. In the case of *Homework 7: Sunrise,* the viewing window has to be set to make the circles visible (or parts of them, perhaps), as well as to make the scales on the *x*- and *y*-axes the same.

- *Where are we in the unit?*

    Once again, have students look at the outline of the unit. They should see that they can check off item 2: "Create the appearance of motion."

2. Introduction of Geometric Transformations

Ask students to turn their attention to the class outline for this unit. They are now going to begin work on the task described in item 3:

> **3. Change the position of an object located in a two-dimensional coordinate system.**

Tell students that the method for doing this will involve a point-by-point function that says, in effect, where each point ends up when it is moved. Tell them that mathematicians often use the word **transformation** for geometric functions, but that students will be looking at a very special type of function, because they don't want to change the shape of objects when they change the positions of individual points.

You may want to focus on a figure such as a triangle and ask what should be true about the new figure compared to the old one. Bring out that the new figure should be *congruent* to the old one.

"In what ways can this triangle be moved without changing its shape?"

Have students picture a triangle as a physical object, such as might be made out of cardboard, and ask how it might be moved without changing its shape. Try to elicit a variety of descriptions. You might get phrases like "move it sideways or up or down," "turn it," and "flip it over."

Tell students that each of these motions is a possibility and that they will look at each case over the course of the unit. (If students don't mention all three basic types of motions, you can tell them that they will learn about the ones they mention "and others.")

> *Note:* A transformation that maintains the shapes of figures is called an *isometry*. (The term is introduced to students in *Homework 24: Flipping Points.*) The term comes from Greek roots meaning "same measure" and refers to the fact that the distance between the images of two points is the same as the distance between the points themselves. Isometries are also called *distance-preserving functions* or *rigid motions*.

As the Cube Turns

Day 8

> Students will be learning about the three basic types of isometries, called *translations, rotations,* and *reflections*. (Translations are introduced today, rotations on Day 11, and reflections in *Homework 24: Flipping Points*.) In *Homework 22: Another Mystery*, students will combine a translation with a rotation, and they may see that the result is actually another rotation. You may want to mention later in the unit (after all three basic types have been discussed) that there are more complex isometries than these, but that all isometries can be made by combining the three basic types.

- *Translations*

    Tell students that one of the basic transformations that they will study is called a **translation,** also known more informally as a **slide.** You can describe a translation as a transformation in which every point is moved the same distance in the same direction.

    To help students get a more precise sense of what this term means, draw a simple diagram like the outline of a house on the board or overhead. Then have a volunteer draw what he or she thinks the result of *translating* or *sliding* this diagram might be. You should get a diagram something like this:

    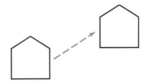

    Tell students that the term *translation* is used to refer both to the *process* of moving the points and to the *result* of the movement. For instance, we might say that we have translated the first house up and to the right. We could also say that the second house is a translation of the first. (The word *translation* has a similar dual usage in its ordinary sense of going from one language to another.)

    > *Note:* The terms *rotation* and *reflection* are also used in these two senses, as is the broader term *transformation.*

- *Numerical description of a translation*

    Tell students that in most of their work on the central unit problem, they will be giving instructions to the calculator in terms of numerical coordinates. Therefore, it will be important for them to be able to describe translations numerically.

Day 8

As an example, have them look at the line segment from (3, 2) to (5, 1). You may want to show this segment on a coordinate grid, as in the accompanying diagram.

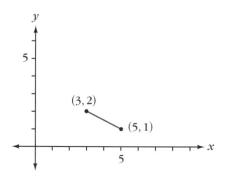

"What other segments are translations of this one?"

Ask each group to find examples of segments that are translations of this initial segment. Because the segment is defined on the calculator by its endpoints, they have to decide what new pairs of endpoints would work.

"What could you do to the original coordinates to get the translated segment?"

Ask students to describe how they could get the coordinates of the new endpoints from the coordinates of the old endpoints. The goal here is to have students see that they can get the endpoints of a translated segment by adding a particular number to the two *x*-coordinates and adding a particular number, possibly different from the first, to the two *y*-coordinates. (But see the note at the end of this subsection for another perspective.)

For instance, if they add 3 to both *x*-coordinates and −4 to both *y*-coordinates, the initial segment is translated to the segment from (6, −2) to (8, −3), as shown in the next diagram.

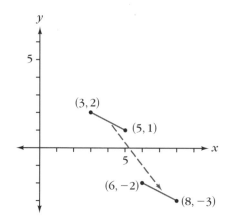

"Why can this translation be represented by the ordered pair (3, −4)?"

Tell students that this translation can be represented by the ordered pair (3, −4), and ask them to explain this representation. They should see that the first number represents how far each point will move in the horizontal direction (to the right or left), and the second how far each point will move in the vertical direction (up or down). You may want to bring out specifically the role of the sign of each number.

62 As the Cube Turns   Interactive Mathematics Program

Day 8

Tell students that in this context, the ordered pair $(3, -4)$ is called a **translation vector,** and is written as $[3 \ -4]$. (Identifying this vector is the first step toward the use of matrices in representing translations. Tomorrow night's homework includes a review of basic ideas about matrices. For now, you may want simply to remind students that a vector is a matrix with only one row or only one column.)

"What does this translation do to the point (a, b)?"

Ask what this translation would do to some general point with coordinates (a, b). Students should see that this translation would move (a, b) to $(a + 3, b - 4)$. Bring out that this corresponds to the sum of the two row vectors $[a \ b]$ and $[3 \ -4]$.

Point out the analogy between representing the translation by the two numbers 3 and -4 and the idea of *components of velocity* that students used in *High Dive*. In both cases, motion in two dimensions is expressed as a combination of vertical and horizontal parts.

> *Note:* The approach just described may not be the way that students actually found the translated segments. For instance, they may have seen that one of the original endpoints, $(5, 1)$, was "two over, one down" from the other original endpoint, $(3, 2)$. They might then have picked one new endpoint arbitrarily and found the other new endpoint by going "two over, one down." If students describe this alternate method, acknowledge that it is correct. But bring out that the "two over, one down" approach does not provide an effective description of the translation overall, because it does not explain what should happen to *every* point.

3. Move That Line!

Tell students that their next programming challenge is to write a program that moves a line segment repeatedly across the screen, using the same translation for each movement. With that introduction, have them begin work on *Move That Line!*

If groups have trouble getting started, tell them to pick endpoints for their initial segment [we will use the earlier example of $(3, 2)$ and $(5, 1)$] and decide where the segment should move to in the first translation. (This is essentially what they are asked to do in Questions 1a and 1b.)

At least some groups likely will initially write programs for Question 2 without using a loop, and this discussion is based on that assumption. For instance, if they use the same initial segment and the same translation as used in the introductory discussion of translations just completed, their plain-language program for Question 2a might look like the program on the next page.

Interactive Mathematics Program
As the Cube Turns 63

Day 8

>Setup program
>
>Draw a line segment from (3, 2) to (5, 1)
>
>Delay
>
>Clear the screen
>
>Draw a line segment from (6, −2) to (8, −3)
>
>Delay
>
>Clear the screen
>
>Draw a line segment from (9, −6) to (11, −7)
>
>Delay
>
>Clear the screen
>
>Draw a line segment from (12, −10) to (14, −11)
>
>Delay
>
>Clear the screen
>
>Draw a line segment from (15, −14) to (17, −15)
>
>Delay
>
>Clear the screen
>
>Draw a line segment from (18, −18) to (20, −19)

Comment: We are assuming here that the setup program includes a "clear screen" instruction.

"How do you find the coordinates of the endpoints of each new segment?"

As you circulate, you may need to help groups as they move on to Question 3. In particular, you may need to help groups clarify the process by which they find the coordinates of the endpoints of each new segment from those for the previous segment. You may also need to give them some help setting up the variables.

For example, they might call one endpoint (P, Q) and the other (R, S). If they start with the segment from (3, 2) to (5, 1), they will need steps in their program that set P to be 3, Q to be 2, R to be 5, and S to be 1. They can then move the segment by changing the values of the variables P, Q, R, and S. (A sample plain-language program is included in tomorrow's discussion of this activity.)

Do not expect all groups to get to this stage today. They can finish this tomorrow, and even then some may only get through Question 3. Question 4 is included for faster groups.

Day 8

You may want to remind students to begin their program with the setup program, so their screen is cleared at the beginning and so their viewing window is set to appropriate values.

Homework 8: Double Dotting

> This homework should help students get a handle on using loop variables as coordinates of lines.

Carol Kaneko and Steve Hansen observe the animation they successfully programmed on their graphing calculators.

Day 8

CLASSWORK

Move That Line!

Your task in this activity is to write a program that will make a line segment move across the screen by repeating the same translation. After you have done this for a single line segment, you will do it for a more complex picture.

1. **a.** Draw a line segment on graph paper.

   b. Choose a translation and draw the result of applying that translation to your line segment.

   c. Apply the same translation to the segment you got in Question 1b, and then apply it again to your new result, and so on, until you have applied the translation five times altogether.

2. **a.** Write a plain-language program that will produce the results you drew on graph paper in Question 1.

   b. Write programming code for the plain-language program of Question 2a.

   c. Enter and test your program from Question 2b, and modify it if necessary.

3. If you didn't use a loop in Question 2, redo those programs using a loop. (*Suggestion:* Use variables for the coordinates of the endpoints, and use steps before you start the loop to give the initial values for these variables. Then think about how the coordinates for each new pair of endpoints are found from the coordinates of the preceding pair, and put steps in your loop to make these changes in the variables.)

4. Once you know how to write a program using a loop to make a single line segment move, make a simple picture out of four or five segments. Then write a program to make that whole picture move through several identical translations.

Double Dotting

This is another homework assignment that presents plain-language calculator programs and asks you to predict their results. In each of Questions 1 and 2, read the program and show on graph paper what should appear on the calculator screen.

1. **Program: DOTS**

    Setup program

    Set A equal to 3

    Set B equal to 2

    Set C equal to 9

    Set D equal to 2

    For P from 1 to 7

    • Draw dots at (A, B) and (C, D)

    • Increase A, B, and D by 1 each

    • Decrease C by 1

    End the P loop

Continued on next page

2. **Program: MOREDOTS**

   Setup program

   Set A equal to 3

   Set B equal to 5

   Set C equal to 3

   Draw a dot at (A, B)

   For P from 1 to 5
   - Decrease A and B by 1
   - Increase C by 1
   - Draw dots at (A, B) and (C, B)

   End the P loop

3. Write a program of your own that uses variables in loops to draw dots.

DAY 9

Students put loops into their translations.

Continuing to Move the Line

Mathematical Topics

• Using loops to create programs with geometric translations

Outline of the Day

In Class

1. Discuss *Homework 8: Double Dotting*
2. Have students complete *Move That Line!* (from Day 8)
3. Discuss *Move That Line!*
   • Focus on the use of a loop to repeat the translation
   • Introduce a "generic" outline for an animation program

At Home

Homework 9: Memories of Matrices

1. Discussion of *Homework 8: Double Dotting*

Have students share the programs they wrote for Question 3 in their groups. While this is happening, choose two students to prepare presentations of Questions 1 and 2. Ask them, as part of their presentations, to go step by step through the program, showing how the values of the variables change at each step so other students can see what is happening. (For instance, they might use boxes to show the value of each variable or make a table with a column for each variable.)

Day 9

2. Completion of *Move That Line!*

As indicated yesterday, we are assuming that at least some groups did Question 2 without the use of loops. By "Question 2" here, we will mean a program written without loops, and by "Question 3," we will mean an approach using loops. If all groups used loops initially when they worked on Question 2, you can go straight to Question 3 when you begin the discussion of the activity.

Let students continue work on *Move That Line!* until most have finished Question 3. As some groups complete Question 3, ask the spade card member of one group to prepare a presentation on Question 3. Also, if possible, identify a group that did Question 2 initially without loops and have the spade card member of that group prepare a presentation. When the presenters are ready, begin the class discussion.

3. Discussion of *Move That Line!*

- **Question 2: Moving the line without a loop**

    Have the student presenting Question 2 go first, to validate this more concrete approach to the problem and to lay a foundation for discussion of the use of the loop for repetition of the translation.

    As noted yesterday, a plain-language program for Question 2a might look like this (although the specific numbers will depend on the initial segment and the translation that groups choose):

    Setup program
    Draw a line segment from (3, 2) to (5, 1)
    Delay
    Clear the screen
    Draw a line segment from (6, −2) to (8, −3)
    Delay
    Clear the screen
    Draw a line segment from (9, −6) to (11, −7)
    Delay
    Clear the screen
    Draw a line segment from (12, −10) to (14, −11)
    Delay
    Clear the screen
    Draw a line segment from (15, −14) to (17, −15)

Day 9

>
> Delay
>
> Clear the screen
>
> Draw a line segment from (18, −18) to (20, −19)

Post the plain-language program for reference during the rest of the discussion.

- **Question 3: Moving the line with a loop**

    Now have a presentation on Question 3. Here are the key elements for writing the program using a loop.

    - Using variables to represent the coordinates of the endpoints
    - Writing instructions by which these coordinates are changed

"How do you get the pairs (18, −18) and (20, −19) from the previous pairs?"

As needed, use the more concrete approach of Question 2 to help students understand the use of the loop. For instance, you can ask how the numbers in the last instruction for Question 2a were obtained from the numbers in the previous instruction for drawing a line segment.

Bring out that the same thing is done to get each of the new *x*-coordinates from the previous *x*-coordinates. For instance, in the example given, students would add 3 to both 15 and 17 to get the new *x*-coordinates 18 and 20. Similarly, the same thing is done to get each new *y*-coordinate from the previous *y*-coordinate. In our example, we are adding −4 (or subtracting 4) in each case. (*Comment:* In anticipation of the discussion of translation matrices, it may be helpful to suggest that students think of the arithmetic as adding −4 rather than subtracting 4.)

Help students to see that the pair of numbers being used to get the new coordinates (in our case, 3 and −4) corresponds to the translation vector [3 −4] used in yesterday's discussion to describe the translation (see the subsection "Numerical description of a translation").

The identification of this translation vector may be helpful for some groups in making the transition in *Move That Line!* to the use of a loop for repeated translation, because it makes explicit the process of obtaining the new coordinates from the old ones.

Also emphasize that the same process is used at each stage to get the next pair of points. Focus on the fact that the repetitious nature of the task makes it suitable for a loop.

Interactive Mathematics Program   As the Cube Turns 71

Day 9

Post for use tomorrow at least one of the programs developed for *Move That Line!* that uses a loop. Here is one possible plain-language program using a loop:

> Setup program
>
> Set P equal to 3, Q equal to 2, R equal to 5, and S equal to 1
>
> For N from 1 to 6
> - Clear the screen
> - Draw the line segment from (P, Q) to (R, S)
> - Delay
> - Add 3 to each of P and R and add −4 to each of Q and S
>
> End the N loop

Reminder: The specific numbers used will vary from group to group, as might the structure of the programs.

- *The generic animation outline*

    Bring out that a loop can be used for many animation programs. At least three things will vary from one animation program to another:

    - How the initial coordinates are set
    - What gets drawn using the coordinates
    - How the coordinates are changed

    As long as there is some way to describe how to get the next set of coordinates from the previous set, this approach will work. For translation, the coordinate change consists of adding the same number to each *x*-coordinate and the same number to each *y*-coordinate.

Post an outline like this one (or something similar developed by the class) as a "generic" animation outline.

> Setup program
>
> Set the initial coordinates
>
> Start the loop
> - Clear the screen
> - Draw the next figure
> - Delay
> - Change the coordinates
>
> End the loop

72 *As the Cube Turns*   Interactive Mathematics Program

Day 9

> *Note:* When this program goes through the loop for the first time, the "next" figure is actually the "first" figure. Alternatively, the program could be written so the first figure is drawn before the loop begins. Also, because the loop in this outline includes an instruction to clear the screen, the setup program need not include one.

Point out to students that this outline is much more of an overview than the plain-language programs they have been writing. Tell them that they will create outlines as part of their work on their animation projects. This will involve an overview like the one just developed, but more specific to their particular projects. You will probably want to discuss this again when the project is assigned on Day 20.

- **Question 4**

  You can use Question 4 as a vehicle for pointing out that the process works even for more complicated diagrams. There would be more pairs of variables, but one would simply assign them initial values and change them the same way as was done with the four variables used previously.

  If time permits, you can have students share their results from Question 4 on the calculator overhead projector.

Homework 9: Memories of Matrices

> This homework should help students remember what they learned about matrices last year. They will begin using matrices tomorrow to work with drawings. This notation will be especially helpful when they work with rotations

Dave Robathan and Sherry Fraser use their calculators' linking capability to quickly exchange lengthy programs.

Interactive Mathematics Program As the Cube Turns 73

Memories of Matrices

Do you remember matrices?

"How could we possibly forget matrices?" you might reply. But just in case, this assignment begins with part of an activity from the Year 3 unit *Meadows or Malls?* to help jog your memory.

It turns out that matrices can be helpful in expressing geometric translations, and you'll be using them to do that in later assignments.

1. (From *Meadows or Malls?*) A matrix could be used to keep track of students' points in a class. Each row could stand for a different student: Clarabell, Freddy, Sally, and Frashy. The first column might be for homework, the second for oral reports, and the third for POWs.

   Suppose the results for the first grading period were represented by this table.

	Homework	Reports	POWs
Clarabell	18	54	30
Freddy	35	23	52
Sally	46	15	60
Frashy	60	60	60

   A matrix representation of this information might look like this.

   $$\begin{bmatrix} 18 & 54 & 30 \\ 35 & 23 & 52 \\ 46 & 15 & 60 \\ 60 & 60 & 60 \end{bmatrix}$$

Continued on next page

Day 9

Here, in table form, are the students' points in each category for the second grading period.

	Homework	Reports	POWs
Clarabell	10	60	0
Freddy	52	35	58
Sally	42	20	48
Frashy	60	60	60

a. Write these second-grading-period scores in a matrix.

b. Figure out each student's total points *in each assignment category* for the two grading periods combined, and write the totals in matrix form.

c. Congratulations! If you completed Question 1b, you have added two matrices. Based on your work, write an equation showing two matrices being added to give the matrix you got in Question 1b.

With this brief review of matrix addition, now look at how matrix addition relates to translations.

2. Suppose you have a diagram of a house set up in a coordinate system like the one shown here.

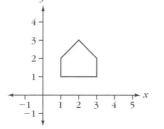

a. Make a matrix A out of the coordinates of the five points that are at the corners of the house, using a separate row of the matrix for each point.

Now suppose you want to translate this diagram, moving it so that the lower left-hand corner of the house ends up at (4, 2).

Continued on next page

138 *As the Cube Turns*

b. Make a matrix B out of the coordinates of the translated house.

c. Find a matrix C so that $A + C = B$. (The matrix C is called a *translation matrix*.)

3. Let's use the letter A for the matrix shown in Question 1. A has 4 rows and 3 columns, so it has 12 entries altogether.

$$A = \begin{bmatrix} 18 & 54 & 30 \\ 35 & 23 & 52 \\ 46 & 15 & 60 \\ 60 & 60 & 60 \end{bmatrix}$$

You will need to be familiar with notation used to represent individual entries of a matrix. Mathematicians use the symbol a_{ij} to refer to the entry in the ith row and j^{th} column (with the lower-case letter a indicating that the entry is from the matrix represented by upper-case A). For example, a_{32} refers to the entry in the third row and second column of matrix A, so a_{32} is 15.

Note: The context makes clear that the numbers 3 and 2 here are separate subscripts, and not the number 32. For larger matrices, you would need to use commas to separate the two subscripts. For instance, $a_{12,34}$ would represent the entry in the 12th row and 34th column.

a. What is a_{23} in the matrix A shown above?

b. Write the entry 30 from matrix A in the form a_{ij}.

c. Write a new matrix B with entries defined by this set of equations:

$$b_{11} = 3, b_{12} = 6, b_{13} = 2$$
$$b_{21} = 5, b_{22} = 1, b_{23} = 3$$

DAY 10

A Matrix Day

Mathematical Topics

- Reviewing matrices
- Using matrices for translations
- Using matrices in graphing calculators

Students learn how to use matrices in programs to create translations.

Special Materials Needed

- A calculator manual for each pair of students

Outline of the Day

In Class

1. Discuss *Homework 9: Memories of Matrices*
   - Post a general principle describing the use of matrices for translations
2. Have students rewrite the *Move That Line!* program using matrices
   - Students can use their manuals to review how to use matrices in their calculators

At Home

Homework 10: Cornering the Cabbage

Discuss With Your Colleagues

And Now They Are Bringing in Matrices!

This unit contains a lot of diverse mathematics and programming ideas, and you and/or your students may be overwhelmed. Now the unit introduces the use of matrices in the programs. Students and teachers without programming experience may be feeling very challenged.

How are you handling all of these concepts? How can you reassure students, especially if you are feeling overwhelmed yourself?

Interactive Mathematics Program — As the Cube Turns

Day 10

1. Discussion of *Homework 9: Memories of Matrices*

Have students discuss the homework in their groups while you take a quick inventory of how well they did on it. If most students have done a good job on the homework, you can go over it quite quickly.

Based on your impression of students' work, you may go directly to Question 2. Otherwise, ask three heart card students to give their answers to Questions 1a through 1c.

- **Question 2**

For Question 2, ask a diamond card student to present the matrix of the initial diagram (Question 2a) and another diamond card student to present the matrix of the translated diagram (Question 2b).

Question 2c is a key element here. You should elicit an explicit equation, which will probably look like this (with perhaps a variation in the order of the rows):

$$\begin{bmatrix} 1 & 1 \\ 1 & 2 \\ 2 & 3 \\ 3 & 2 \\ 3 & 1 \end{bmatrix} + \begin{bmatrix} 3 & 1 \\ 3 & 1 \\ 3 & 1 \\ 3 & 1 \\ 3 & 1 \end{bmatrix} = \begin{bmatrix} 4 & 2 \\ 4 & 3 \\ 5 & 4 \\ 6 & 3 \\ 6 & 2 \end{bmatrix}$$

Identify the matrix C as the **translation matrix** and bring out that its rows are all the same. Bring out that C simply consists of many copies of the *translation vector* associated with moving three units to the right and one unit up.

Post a rule similar to this one to describe translations.

To translate a matrix of points a units in the x-direction and b units in the y-direction, add the

matrix $\begin{bmatrix} a & b \\ \cdot & \cdot \\ \cdot & \cdot \\ \cdot & \cdot \\ a & b \end{bmatrix}$ **to the matrix of points.**

- **Question 3**

Finally, ask club card students to answer the different parts of Question 3. The point of this question is to acquaint them with standard notation that is similar to the notation used by their calculators. Go over the specific notation used by students' graphing calculators for matrix entries, which will probably look something like $a(i, j)$ rather than a_{ij}.

78 *As the Cube Turns* Interactive Mathematics Program

Day 10

2. Moving the Line with Matrices

"How might we use matrices in this program?"

Turn to a posted program for *Move That Line!* (one that uses a loop) and ask students how the ideas about matrices from last night's homework could be used in the program.

> Here is the sample program from yesterday's discussion:
>
> Setup program
>
> Set P equal to 3, Q equal to 2, R equal to 5, and S equal to 1
>
> For N from 1 to 6
> - Clear the screen
> - Draw the line segment from (P, Q) to (R, S)
> - Delay
> - Add 3 to each of P and R and add −4 to each of Q and S
>
> End the N loop
>
> If students' programs for *Move That Line!* were quite different from this, then you may need to spend some time developing a program of this type so that the adaptation to matrices can be done easily.

Bring out that what is involved is merely a change in notation. (This isn't meant to minimize the difficulty students may have with the change.) Instead of some collection of individual variables for the coordinates (such as P, Q, R, and S), we use one variable—a matrix. Individual entries of that matrix represent the coordinates, with each row of the matrix containing the two coordinates for a different point. Instructions involving the coordinates are replaced by similar matrix instructions. For instance, the initial line giving the values for P, Q, R, and S is replaced by a single instruction defining a matrix.

A complete plain-language program might look like this:

Setup program

Let A be the matrix $\begin{bmatrix} 3 & 2 \\ 5 & 1 \end{bmatrix}$

Let B be the matrix $\begin{bmatrix} 3 & -4 \\ 3 & -4 \end{bmatrix}$

For N from 1 to 6
- Clear the screen
- Draw the line segment from (a_{11}, a_{12}) to (a_{21}, a_{22})
- Delay
- Replace matrix A by the sum A + B

End the N loop

Interactive Mathematics Program · · · As the Cube Turns 79

Day 10

This use of matrices may seem like an added complication to students, but if there were many points, it would actually represent a simplification. Also, the use of a matrix to represent the two points will provide a convenient mechanism for modifying the points. (In other examples, we will work with more points, and so have more rows. Also, later in the unit, we will have points in 3-space, so we will need three columns.)

Note: You should make an overhead transparency or other written record of this program, and save it to refer to on Day 18.

- *Writing the matrix program in code*

"What's the programming code for this plain-language program?"

Have students work in groups to create programming code for this plain-language program (or for their own analogous version from their work on *Move That Line!*). In particular, this will require them to review how to define a matrix on their calculators and learn how to refer to individual entries from a matrix.

They can probably figure this out using their calculator manuals. If some groups seem lost, ask for a volunteer from a successful group to tutor them. If no groups figure it out, you may need to pull the class together and go over the procedure for your graphing calculator, using the calculator overhead projector.

Have each student enter the program into the calculator to see how it works. Students should write out their matrix programs on paper so they have them to refer to on Day 18.

Homework 10: Cornering the Cabbage

> Tonight's homework assignment will facilitate the discovery tomorrow that the area of a triangle is half the product of the lengths of two sides times the sine of the included angle. Students will later use that formula to get an expression for the sine of the sum of two angles. They will need the angle sum formula for calculating rotations of points.

Tell students that although this assignment seems unrelated to the material at hand, it will contribute to their ability to do item 3 of their unit outline. You may want to suggest that they use something like straws or toothpicks at home to help visualize the problem.

HOMEWORK 10

Cornering the Cabbage

Did you ever read about Peter Rabbit and his siblings, Flopsie, Mopsie, and Cottontail?

Actually, it doesn't matter whether you did or not. All you need to know is that Peter used to steal cabbages from a local farmer, Mr. McGregor. But now Peter and his sibling Flopsie are working for Mr. McGregor.

Peter and Flopsie usually got paid in cabbages, but Mr. McGregor decided that because they had been working so hard, he'd pay them by giving them some land. That would allow them to grow some of their own cabbages. Peter and Flopsie thought this was a great idea.

Two poles were lying nearby, and Mr. McGregor gave one to Peter and one to Flopsie. He said they could hold them together at the ends and spread them out at any angle they wanted. Then he'd connect the two other ends, forming a triangle. They'd get all the land inside the triangle.

Continued on next page

It turned out Peter's pole was approximately 2 meters long and Flopsie's was approximately 3 meters long. Of course, they wanted the most land they could get, and they weren't sure what angle to choose, or even if it mattered. Can you help them?

1. **a.** Figure out what the area would be if the angle formed by the two poles was 25°.

   b. Figure out what the area would be if the angle was 100°.

2. Pick two other angles to try. Figure out what area Peter and Flopsie would get if they used each of those angles.

3. Develop a general formula for the area of the triangle. Use t and u for the lengths of the two given sides and θ for the angle they form, as in the diagram at the right.

4. Determine what value for the angle θ would give Peter and Flopsie the largest area, and justify your answer.

Day 11

Days 11-19

Rotating in Two Dimensions

This page in the student book introduces Days 11 through 19.

Jeff Tung, Ken Hoffman, and Audrey Rae use coordinates to help them work in two and three dimensions.

The next step in turning the cube in three dimensions is learning to turn things in two dimensions. To accomplish this, you will need to develop some trigonometric formulas. You probably won't be surprised to learn that with rotations, as with translations, matrices can also prove very useful.

DAY 11

Students develop a new formula for the area of a triangle.

An Area Formula

Mathematical Topics

- Developing a formula for area using trigonometry
- Introducing rotations as a type of geometric transformation

Outline of the Day

In Class

1. Discuss *Homework 10: Cornering the Cabbage*
   - Have students develop a generalization based on specific examples
   - Have students extend the generalization to obtuse angles
   - Have students justify the conclusion that the area is maximized when θ is 90°
2. Discuss where the class is in the unit
   - Identify two aspects of item 3—translations and rotations
   - Students should see that they have completed part of item 3 of the unit outline
3. Introduce rotations
   - Illustrate the idea of a rotation with a specific example

At Home

Homework 11: *Goin' Round the Origin*

Interactive Mathematics Program

As the Cube Turns **85**

Day 11

1. Discussion of *Homework 10: Cornering the Cabbage*

> Although the apparent goal in the problem is to maximize the area, the purpose of the problem in the unit is to get the general formula $A = \frac{1}{2} tu \sin \theta$. So be sure that this generalization gets primary attention in the discussion.

Ask students to compare what they found for different angles in their groups. Begin the class discussion by having someone explain the reasoning used to answer each part of Question 1. Students will probably use diagrams like these:

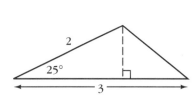
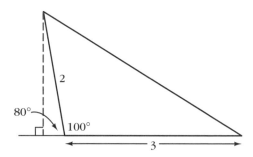

Possible diagram for Question 1a   **Possible diagram for Question 1b**

> *Note:* In both cases here, we show a diagram that uses the side of length 3 as the base. Students might also set up either of the diagrams using the side of length 2 as the base. For Question 1a, that approach proves a bit more difficult to work with than the diagram here because it leads to an altitude that is outside the triangle. For Question 1b, using either of the sides that form the 100° angle as the base will lead to an altitude that is outside the triangle.

"Why should sin 80° and sin 100° give the same answer?"

Students should see that the length of the altitude in the first triangle is 2 sin 25° and that the length of the altitude in the second triangle can be written as either 2 sin 100° or 2 sin 80°.

You should certainly accept both the 100° and the 80° answers, because both are correct. However, for the purposes of generalizing these examples, it's important that students see that the altitude of the second triangle can be written as 2 sin 100°. Use the occasion to review the general definition of the sine function, having students explain why sin 80° and sin 100° are equal. Try to elicit explanations both in terms of the formal definition of the sine function and in terms of the Ferris wheel metaphor (from the Year 4 unit *High Dive*).

Students should then be able to explain how to use the formula $\frac{1}{2}bh$ to get an area of $\frac{1}{2} \cdot 3 \cdot 2 \sin 25°$ for the first triangle and an area of $\frac{1}{2} \cdot 3 \cdot 2 \sin 100°$ for the second triangle. (You may want to review the use of the formula $\frac{1}{2}bh$ for the case of an obtuse triangle.) These expressions simplify to $3 \sin 25°$ and $3 \sin 100°$, which give areas of approximately 1.27 m² and 2.95 m², respectively.

If students were clear on the examples from Question 1, you can probably skip discussion of Question 2.

• Question 3

"What did you do when the angle was 25°?"

Then have a volunteer present Question 3. If no one came up with the general formula $\frac{1}{2}tu \sin \theta$, focus students on the diagram in the homework showing an acute angle for θ. Bring out that the first part of Question 1 was the case where $t = 2$, $u = 3$, and $\theta = 25°$. By having students follow their reasoning in that example, they should be able to come up with the generalization.

Tell students that this rule for calculating the area of the triangle will come in handy soon. Post the general rule, accompanied by an appropriate diagram.

The area of a triangle is equal to half the product of the length of two sides times the sine of the angle between them.

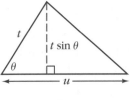

Area = $\frac{1}{2} tu \sin \theta$

• Obtuse angles

"Does the formula work for angles greater than 90°?"

If students seemed unclear earlier on the use of the expression $2 \sin 100°$ for the altitude for the second example in Question 1, ask whether the area formula $\frac{1}{2}tu \sin \theta$ is valid for the case where the angle θ between the given sides is more than 90°.

Day 11

By using a diagram like the one below and referring to their work with Ferris wheels in *High Dive,* students should be able to justify the fact that the altitude in this diagram is still equal to $t \sin \theta$. For instance, if they view the triangle in a coordinate system with the vertex of angle θ at the origin, then the length of the altitude is the y-coordinate of the other end of the side of length t. Thus, this length is $t \sin \theta$.

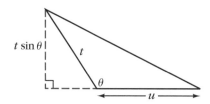

- Question 4: Maximizing the area

Have students discuss Question 4 concerning maximizing the area. Based on the formula, the diagram, and their intuition, they should see that the best choice for Peter and Flopsie is to form a 90° angle with their sticks.

"How can you justify your answer?"

Ask volunteers to present their justifications for this conclusion. They may see geometrically, from a diagram like the last one, that the altitude will be the biggest when the side of length t is vertical. They may also use the fact that the sine function has a maximum value of 1 when the angle is 90°.

"Why does the sine function have its maximum at 90°?"

Ask students to explain why the maximum value for sine function occurs when the angle is 90°. They might do this in terms of the formal definition, in terms of the Ferris wheel model, or with a graph.

2. Where Are We in the Unit?

Have students look at the unit outline from Day 1. They may think they are ready to check off item 3: "Change the position of an object located in a two-dimensional coordinate system." Tell them that they have done part of this—changing the position using a translation—but that their next task involves a different type of position change—a rotation. (Last night's homework is headed in that direction, although it might seem quite unrelated.)

You may want to modify the outline by indicating two subitems for item 3:

> **3. Change the position of an object located in a two-dimensional coordinate system.**
> - **Translate the object.**
> - **Rotate the object.**

You can then check off the first of these subitems.

88 *As the Cube Turns*

Day 11

• *Translations in three dimensions*

"How would you represent translations in three dimensions?"

Ask students how they would represent translations in three dimensions. You need not get into a detailed discussion here. Bring out that the basic ideas are the same and that students would simply need to use a 1-by-3 row vector as the translation vector in place of a 1-by-2 row vector.

Tell students that when they get to three dimensions (item 5 of the unit outline), they will work only with rotations (and not with translations). You might want to emend the outline to reflect this, so the item looks like this:

> **5. Change the position of an object located in a three-dimensional coordinate system.**
>
> • **Rotate the object.**

3. Rotations: An Introduction

Tell students it is now time to examine how to rotate an object in two dimensions. Also tell them that they will see over the next few days how the area formula from last night's assignment fits into this task.

"Where would this point end up if we rotated it 20° around the origin?"

Draw coordinate axes on the overhead projector. Then ask students to identify a specific point in the plane (other than the origin) and to explain what it would mean to rotate that point 20° counterclockwise around the origin. Help them to articulate that rotating the point around the origin means that the distance from the point to the origin will stay fixed and that the counterclockwise angle from the positive *x*-axis will increase by 20°. Be sure to draw in the radii to the origin, so students realize exactly what stays fixed.

For example, a 20° counterclockwise rotation of the point labeled *P* in the next diagram would put it at the position labeled *Q*.

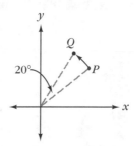

Tell students that they will be looking for ways to get the coordinates of *Q* in terms of the coordinates of *P* and the size of the angle of rotation.

Homework 11: Goin' Round the Origin

> This assignment is a first step toward developing a formula for rotation around the origin.

Interactive Mathematics Program   As the Cube Turns **89**

Goin' Round the Origin

Being able to rotate pictures around the origin will be useful when trying to create interesting graphics. And, of course, this will be important in turning a cube.

This assignment begins the task by examining a specific problem and its generalization.

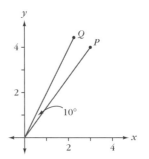

1. Consider the point P in the accompanying diagram, whose coordinates are $(3, 4)$. Suppose this point is rotated $10°$ counterclockwise around the origin to the position shown as point Q.

What are the coordinates of Q? (*Hint:* You may find trigonometry useful.)

2. Next, consider the point R with coordinates $(2, 5)$. As in Question 1, rotate this point $10°$ counterclockwise around the origin. What are the coordinates of its new location?

3. Now generalize Questions 1 and 2. Begin with a point with coordinates (x, y). Rotate that point $10°$ counterclockwise around the origin. Develop a formula or procedure for getting the coordinates of the new point in terms of x and y.

DAY 12

Goin' Round the Origin

Students find formulas for rotations that don't always work.

Mathematical Topics

- Finding coordinate formulas for rotation in the plane
- Reviewing polar coordinates and their relationship to rectangular coordinates

Outline of the Day

In Class

1. Discuss *Homework 11: Goin' Round the Origin*
   - Have students share formulas, which will probably involve the inverse tangent function
   - Bring out that formulas using the inverse tangent function don't work in all quadrants
2. Review polar coordinates
   - Review and post the basic relationships $x = r \cos \theta$ and $y = r \sin \theta$
   - Have students express the rotation in terms of polar coordinates

At Home

Homework 12: Double Trouble

Day 12

1. Discussion of *Homework 11: Goin' Round the Origin*

Have a spade card student present Question 1. We describe one likely approach here.

Begin by finding the angle shown as θ in the diagram, perhaps using the condition $\tan \theta = \frac{4}{3}$, so $\theta = \tan^{-1}\left(\frac{4}{3}\right)$. This gives $\theta \approx 53°$.

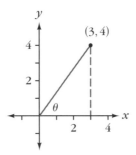

Then, using (a, b) to represent the new coordinates as in the next diagram, add 10° to θ, getting approximately 63° for the angle shown. Use the fact that the length of the segment from the origin to (a, b) is the same as the length of the segment from the origin to $(3, 4)$. This length is 5, as can be found by the Pythagorean theorem.

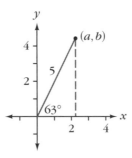

Students can then find the coordinates a and b, either from the right triangle or based on the general definition of the sine and cosine functions. Either approach gives $a \approx 5 \cos 63° \approx 2.26$ and $b \approx 5 \sin 63° \approx 4.46$.

Bring out that these are reasonable values, based on the diagram. (*Note:* You may want to save these numerical values for comparison when students develop general formulas for rotation on Days 14 and 15.)

- Question 2

Next, ask for another spade card student to present Question 2. Use your judgment as to how much detail your class needs in this presentation. (You might ask students how much detail they want.) The coordinates of the new point are approximately (1.10, 5.27). (*Note:* You may also want to save this result for comparison when the general formula is developed.)

- **Question 3**

Ask for a volunteer to present the generalization in Question 3. Students should be able to mimic the specific examples to articulate that the new *x*-coordinate, which we'll call x' (read as *x prime*), is given by this equation:

$$x' = \sqrt{x^2 + y^2} \cos\left[\tan^{-1}\left(\frac{y}{x}\right) + 10°\right]$$

Similarly, they should find that the new *y*-coordinate, y', is given by this equation:

$$y' = \sqrt{x^2 + y^2} \sin\left[\tan^{-1}\left(\frac{y}{x}\right) + 10°\right]$$

Post these formulas, for comparison with another version tomorrow. Students will probably think that these formulas are a mess.

> *Note:* If someone used something like $\sin^{-1}\left(\frac{y}{\sqrt{x^2+y^2}}\right)$ instead of $\tan^{-1}\left(\frac{y}{x}\right)$, you can look at that as well and see that it's even messier.

- *Trouble in the third quadrant*

"What do you get when you apply these formulas to try to rotate $(-3, -4)$?"

Next, have students try out these formulas with the point $(-3, -4)$. They should see that the formulas don't work, because they yield the same answer for the point $(-3, -4)$ as they do for the point $(3, 4)$.

"Why don't the formulas work?"

Let students mull this over for a bit. Ask if they can figure out what went wrong. If necessary, you can remind them of their work in *High Dive*, where they saw the issue of the ambiguity of the inverse trigonometric functions (for instance, in *Homework 7: More Beach Adventures*). In particular, they saw that for any number between -1 and 1, there are infinitely many angles whose sine or cosine is that number. A similar situation happens with the tangent function. Therefore, using $\tan^{-1}\left(\frac{y}{x}\right)$ for getting the new angle will not work. Something else is needed. Tell students that tonight's homework will get them a bit closer to a better set of formulas for x' and y'.

2. Review of Polar Coordinates

Tell the class that there is hope for the development of simpler formulas and that polar coordinates can lead the way. Take this opportunity to review polar coordinates as needed, especially the formulas for finding rectangular coordinates from polar coordinates.

Day 12

"How do you find the rectangular coordinates from the polar coordinates?"

Using a diagram like this one, ask how to express the rectangular coordinates x and y in terms of the polar coordinates r and θ.

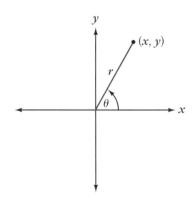

Students should be able to use the right triangle, if necessary, to get the two general formulas:

$$x = r \cos \theta \text{ and } y = r \sin \theta$$

Post these formulas (if they aren't still posted from *High Dive*).

"Will these formulas work for all values of θ?"

Ask whether these formulas will work for all values of θ, that is, for all quadrants. This is an especially important question in the aftermath of the failure of the prior formulas involving the inverse tangent to generalize.

Use your judgment about how much of a review of the ideas from *High Dive* is appropriate here. As needed, remind the class that the sine and cosine functions are *defined* in general by the formulas $\cos \theta = \left(\frac{x}{r}\right)$ and $\sin \theta = \left(\frac{y}{r}\right)$, so the formulas for x and y in terms of r and θ hold for all angles.

• Rotation formulas using polar coordinates

"How can you rewrite the formulas for rotations using polar coordinates?"

Ask students how the flawed rotation formulas developed earlier can be rewritten using polar coordinates. They should recognize the expression $\sqrt{x^2 + y^2}$ as equal to r and the expression $\tan^{-1}\left(\frac{y}{x}\right)$ as equal to θ. (In fact, what students need in the formula is *always* θ, and θ is only *sometimes* equal to $\tan^{-1}\left(\frac{y}{x}\right)$.) Thus, they should see that if the original point has polar coordinates (r, θ), then the result of rotating 10° is the point (x', y'), where

$$x' = r \cos(\theta + 10°)$$
$$y' = r \sin(\theta + 10°)$$

and where $r = \sqrt{x^2 + y^2}$.

Post this new pair of formulas for x' and y' for comparison with the earlier versions.

Day 12

"Will these new formulas work in all quadrants?"

Ask students whether these new formulas will work for all values of θ. They should see that they do, but that to use these new formulas productively, they seem to need two things.

- A good expression for θ in terms of x and y
- A way to handle the sine and the cosine of a sum

Tell students that with the help of tonight's homework, they will develop a way to deal with the second of these items and at the same time eliminate the need for the first. That will allow them to avoid the use of the inverse tangent function, which doesn't necessarily give the desired value outside the first quadrant.

Homework 12: Double Trouble

Tonight's homework is intended to ease students into discovering the formula for the sine of a sum, which will be the key for expressing rotations.

Julia Au, Stephanie Skangos, and DeAnna DelCarlo set up a coordinate system in which to analyze the effect of a rotation around the origin.

Interactive Mathematics Program — As the Cube Turns

Double Trouble

Do you remember Woody from the *Shadows* unit in Year 1? As you may recall, Woody was very fond of measuring trees. He was especially happy to learn how to use trigonometry so he could avoid climbing trees.

One time, Woody leaned a ladder between the branches so that its tip just reached the top of a tree. Next, he measured the angle that the ladder made with the ground. He wrote down the angle and headed home to do the computation to find the tree's height. (He knew the length of the ladder.)

When he got home, he grabbed his scientific calculator, punched in sin 40°, and got 0.643. Then he punched in cos 40°, just in case he might want to know how far the foot of the ladder had been from the tree. He found that cos 40° was about 0.766.

When he did the calculation, he realized that his answer seemed way off. He ran back to the scene, and measured the angle again. Lo and behold, it was really 80°!

When Woody returned home, he couldn't find his calculator anywhere. His friend Elmer suggested that Woody simply multiply sin 40° by 2 to get sin 80°.

1. Is Elmer's suggestion correct? That is, is sin 80° equal to 2 sin 40°?

2. How could you decide if he was right if you had no way to look up sin 80°?

Continued on next page

In *Homework 10: Cornering the Cabbage*, you saw that the area of a triangle such as the one shown here is given by the expression $\frac{1}{2}tu \sin \theta$. Use this formula in Questions 3a and 3b.

3. The first diagram below shows an isosceles triangle with two sides of length x and an angle of 80° formed by those sides. The second diagram shows the same triangle with the altitude drawn. The altitude has length z and splits the original triangle into two smaller triangles.

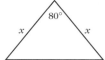 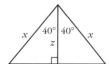

   a. Find the area of the original triangle (in terms of x).

   b. Find the area of each of the two small triangles (in terms of x and z).

4. a. Use the fact that the large triangle is made up of the two smaller triangles to get an expression for sin 80°. This expression may involve sin 40° and both x and z.

   b. Work with your answer to Question 4a to get an expression for sin 80° that involves neither x nor z, but may involve both sin 40° and cos 40°.

DAY 13 The Sine of a Sum

Students begin to develop the formula for the sine of a sum.

Mathematical Topics

• Discovering and justifying a formula for the sine of the sum of two angles

Outline of the Day

In Class

1. Form new random groups
2. Discuss *Homework 12: Double Trouble*
   • Emphasize that sin 80° is not equal to twice sin 40°
   • Use area to develop the relationship sin 80° = 2 sin 40° cos 40°
3. *The Sine of a Sum*
   • Students work to develop the formula for the sine of a sum
   • The activity will be discussed on Day 14

At Home

Homework 13: A Broken Button

1. Forming New Groups

This is an excellent time to form new random groups. Follow the procedure described in the IMP *Teaching Handbook,* and record the members of each group and the suit for each student.

2. Discussion of *Homework 12: Double Trouble*

Ask a heart card student to summarize Woody's problem. His dilemma is that he knows the values of both sin 40° and cos 40°, but he wants to find the value of sin 80°.

"Can Woody simply multiply sin 40° by 2 to get sin 80°?"

Then turn to Question 1, asking the class whether Woody can simply multiply sin 40° by 2. Presumably, students found that sin 80° and 2 sin 40° are not equal, probably by evaluating both expressions on a calculator.

Interactive Mathematics Program   As the Cube Turns 99

Day 13

"How can you show that sin 80° and 2 sin 40° are unequal without a calculator?"

Then move on to Question 2, asking how students might show that sin 80° and 2 sin 40° are unequal without a calculator.

"What is 2 sin 40°? Why can't sin 80° be equal to this value?"

If no one has any suggestions, you might have students find 2 sin 40° (using the fact that sin 40° is .643). Bring out that because 2 sin 40° is more than 1, it can't be the sine of any angle.

"Are there angles whose sine you know without the use of a calculator?"

Also ask if there are angles whose sine students know without the use of a calculator. If necessary, suggest that they think about the case of 90° and 180°. They should see that sin 180° is not equal to 2 sin 90°.

Bring out that the condition "sin 180° ≠ 2 sin 90°" does not prove that sin 80° is not equal to 2 sin 40°, but it does show that the equation sin 2A = 2 sin A does not hold for all angles. (And as students have just seen, it cannot hold if sin A is greater than .5.)

"How can you use the Ferris wheel model to see why sin 80° ≠ 2 sin 40°?"

Ask if anyone can use the Ferris wheel model to see why sin 80° ≠ 2 sin 40°. Try to elicit the explanation that going 80° around the Ferris wheel (from the 3 o'clock position) does not place the rider twice as far above the center as going 40° around the Ferris wheel.

You may want to ask students to check other examples on their calculators. You can also have them see from the graph of the sine function that doubling the x-value does not lead to doubling the y-value.

- **Question 3**

    Next, ask other heart card students for expressions for the areas of the large triangle and the small triangles. They will probably give these equations:

    $$\text{Area of large triangle} = \frac{1}{2}x^2 \sin 80°$$

    $$\text{Area of each small triangle} = \frac{1}{2}xz \sin 40°$$

- **Question 4**

    Ask for a volunteer to answer Question 4a. (If no one got an answer, have students work on this in their groups. As a hint, ask them how the areas of the triangles are related.) From the fact that the area of the large triangle is the sum of the areas of the two small triangles, they should get the equation

    $$\frac{1}{2}x^2 \sin 80° = 2 \cdot \frac{1}{2}xz \sin 40°$$

    Multiplying both sides by 2, dividing both sides by x^2, and simplifying should lead students to the equation

    $$\sin 80° = 2 \cdot \frac{z}{x} \cdot \sin 40°$$

Day 13

"What does the ratio $\frac{z}{x}$ represent within the small right triangle?"

Next, turn to Question 4b. Again, if necessary, let students work in their groups to "get rid of x and z" or to "bring cos 40° into the equation." If they need a hint, remind them that Woody knows cos 40°, or else ask what the ratio $\frac{z}{x}$ represents within the small right triangle. This should lead them to the formula

$$\sin 80° = 2 \cos 40° \sin 40°$$

You can connect this formula with work so far on rotations by writing this as

$$\sin (40° + 40°) = 2 \cos 40° \sin 40°$$

Bring out, if necessary, that this equation involves finding the sine of a sum of two angles, as do the rotation formulas using polar coordinates, which involve $\sin (\theta + 10°)$.

Let students speculate briefly about whether the last equation is something peculiar to 40° or whether it generalizes. Then tell them that their next activity is to generalize this idea to a case in which the two angles being added are not equal.

3. *The Sine of a Sum*

Let students begin work immediately on *The Sine of a Sum*. Discussion is scheduled for tomorrow. As groups conclude their work, ask one or two groups to prepare presentations.

Homework 13: A Broken Button

This homework will pave the way for finding a formula for the cosine of a sum on Day 15.

Day 13

The Sine of a Sum

You have seen that when a point is rotated 10° around the origin, the rectangular coordinates for its new location involve the expressions cos (θ + 10°) and sin (θ + 10°).

It would be helpful if you knew how to find the sine or cosine of two angles added together, using their individual sines and cosines. In this activity, you'll begin with sine.

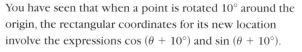

Your task is to develop a formula for sin (A + B), where A and B are two angles. You can use sin A, cos A, sin B, and cos B in your formula.

Suggestion: Modify your work on *Homework 12: Double Trouble* using a diagram like the one shown here.

A Broken Button

Elmer's phone rang. He knew it would be Woody even before he picked it up. Ever since Woody had lost his calculator, he'd been pestering Elmer.

This time, Woody wanted to know the value of cos 50°. But when Elmer went to find out on his own calculator, he discovered that the cosine key was broken.

Fortunately, the sine key was working. Elmer offered to give Woody the value of sin 50° if that would help. Woody thanked him, wrote down the value of sin 50°, hung up, and proceeded to find cos 50° using the Pythagorean identity, $\sin^2 x + \cos^2 x = 1$.

1. Show how Woody might have used the Pythagorean identity and the numerical value of sin 50° to determine the value of cos 50°.

Later, Elmer called to tell Woody that he'd found an easy, no-computation way to get cos 50°, by using a right triangle and simply finding the sine of a different angle.

2. How do you think Elmer's method worked?

DAY 14

Conclusion of *The Sine of a Sum*

Students continue to work with area to develop a formula for the sine of a sum.

Mathematical Topics

- Developing relationships between sine and cosine
- Completing development of the formula for the sine of the sum of two angles

Outline of the Day

In Class

1. Discuss *Homework 13: A Broken Button*
   - Elicit a formula for cosine in terms of sine using the Pythagorean identity
   - Elicit the formula $\cos \theta = \sin (90° - \theta)$
2. Discuss *The Sine of a Sum* (from Day 13)
   - Develop and post the formula $\sin (A + B) = \sin A \cos B + \cos A \sin B$
   - Have students verify the formula using specific cases

At Home

Homework 14: *Oh, Say What You Can See*

1. Discussion of *Homework 13: A Broken Button*

Give students a few minutes to share ideas on Questions 1 and 2 within their groups. Then have diamond card students from two groups present results on the two problems.

Interactive Mathematics Program   As the Cube Turns 105

Day 14

- **Question 1**

    Using the Pythagorean identity, one would get $\sin^2 50° + \cos^2 50° = 1$. Because Woody knows the value of $\sin 50°$ (approximately .766), he can substitute and get $.766^2 + \cos^2 50° = 1$ which means $\cos^2 50° \approx .413$ and $\cos 50° \approx .643$.

    > *Note:* If many students seem to be vague about the Pythagorean identity, then you might take a few minutes here to have the class review where it comes from and why it's true for all angles.

- **Question 2**

    Students briefly reviewed the identity $\sin \theta = \cos (90° - \theta)$ in *High Dive*, so they may have seen immediately that $\cos 50°$ is the same as $\sin 40°$. If not, or if you feel that a full discussion is needed, you can have students explain their reasoning using a right triangle like the one shown here.

    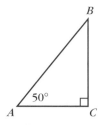

    They should see that on the one hand, $\cos 50°$ is equal, by definition, to the ratio $\frac{AC}{AB}$. On the other hand, this same ratio is also the sine of $\angle CBA$, and students should see that $\angle CBA = 40°$, so $\cos 50° = \sin 40°$.

- **Generalizing the result**

"What is the relationship between 40° and 50°? What are such angles called?"

Ask students how 40° and 50° are related to each other, and review that a pair of angles whose sum is 90° are called *complementary angles*.

"Can you generalize the formula you got for 50°?"

Then ask them to generalize the equation $\cos 50° = \sin 40°$. They should come up with the formula

$$\cos \theta = \sin (90° - \theta)$$

and see that this makes sense for any right triangle with an acute angle of θ.

> *Note:* This is a slight variation on the identity $\sin \theta = \cos (90° - \theta)$, which was reviewed in *High Dive*, and both forms will be needed in tomorrow's work.

106 *As the Cube Turns*

Day 14

"Does this relationship hold for nonacute angles?"

Ask students whether the identity $\cos \theta = \sin (90° - \theta)$ also holds for nonacute angles, where the right-triangle reasoning falls apart. They may choose to test this with values of θ from various quadrants, or they may try to reason based on the graphs of the functions. (The latter idea is pursued in tonight's homework.)

Comment: It is important for the unit problem that students accept this general formula. It is not essential that they know the details of a proof.

If students did not already use the formula $\sin \theta = \cos (90° - \theta)$, ask if there is a formula similar to $\cos \theta = \sin (90° - \theta)$ that Woody could have used if the buttons had been reversed—that is, if his sine key had been broken, and he wanted to find, say, sin 50° using his cosine key. This should lead them to develop the variation.

Post both of these formulas, because students will need to use them tomorrow.

$$\cos \theta = \sin (90° - \theta)$$
$$\sin \theta = \cos (90° - \theta)$$

This is a good opportunity to suggest that students keep a list of trigonometric identities. They will be using known identities to develop new ones, so it will be helpful to have a single place where they can find this material.

2. Discussion of *The Sine of a Sum*

"Why might you want a formula for the sine of a sum?"

Before beginning the presentations on *The Sine of a Sum*, ask students why they might want a formula for the sine of a sum. If a hint is needed, point to the posted formulas for rotations that involve the expressions $\cos (\theta + 10°)$ and $\sin (\theta + 10°)$.

• Developing the formula

If students use the diagram provided in the activity (and shown below), they should see that the area of the large triangle is $\frac{1}{2}xy \sin (A + B)$ and that the two smaller triangles have areas $\frac{1}{2}xz \sin A$ and $\frac{1}{2}yz \sin B$ respectively.

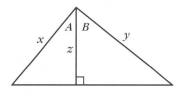

This should lead them to the equation

$$\tfrac{1}{2}xy \sin (A + B) = \tfrac{1}{2}xz \sin A + \tfrac{1}{2}yz \sin B$$

Interactive Mathematics Program   As the Cube Turns   107

One approach from there is to multiply both sides by 2, divide by xy, and simplify. (If a hint is needed, you might refer students to their similar work on *Homework 12: Double Trouble*.) This gives

$$\sin(A+B) = \frac{z}{y}\sin A + \frac{z}{x}\sin B$$

But the ratios $\frac{z}{y}$ and $\frac{z}{x}$ are simply $\cos B$ and $\cos A$ respectively, so the equation becomes

$$\sin(A+B) = \cos B \sin A + \cos A \sin B$$

Post this formula prominently and label it "Sine of a Sum."

Point out that the argument used here only proves the relationship for acute angles. But also tell students that the formula holds for all angles, and let them spend a few minutes verifying with their calculators that this formula holds for angles in all quadrants. (See the supplemental problems *Moving to the Second Quadrant* and *Sums for All Quadrants* for work on proving this equation more generally.)

- *Verifying special cases*

    You can have students check some special cases, such as $B = 0°, 90°,$ and $180°$. Help students see (using the specific values of sine and cosine for these angles) that in these cases, the sine-of-a-sum formula simplifies to these equations:

    - $\sin A = \sin A$
    - $\sin(A + 90°) = \cos A$
    - $\sin(A + 180°) = -\sin A$

    The first is, of course, true for all values of A. You can suggest that students use the graphs of the sine and cosine function to develop an explanation for the second result. (They might also look at the supplemental problem *A Shift in Sine* from *High Dive*, which involves a slight variation on this identity.)

    The third of these identities is considered in the supplemental problem *Adding 180°*. Students may be able to give a good explanation at this point in terms of the graph or the Ferris wheel model.

    Also ask what the sine-of-a-sum formula says if B is equal to A. Bring out that it then becomes the "double-angle" formula

    $$\sin 2A = 2 \cos A \sin A$$

    and that this is consistent with the result students got in *Homework 12: Double Trouble*, namely

    $$\sin 80° = 2 \cos 40° \sin 40°$$

Day 14

Homework 14: Oh, Say What You Can See

This assignment will remind students of some other formulas relating sine and cosine that they saw in *High Dive*.

A copy of the graph showing the two functions is included in Appendix C.

David Granados discusses the development of the formula for the sine of the sum of two angles.

Interactive Mathematics Program · As the Cube Turns 109

Oh, Say What You Can See

In recent work, you may have used the trigonometric identity $\cos \theta = \sin (90° - \theta)$. (*Reminder:* An *identity* is an equation that holds true no matter what values are substituted for the variables.)

Your task in this activity is to give specific examples illustrating this and other identities, and then to explain each identity in one or more of these two ways:

- Using the situation of the Ferris wheel (from *High Dive*)
- Using the graphs of the equations $y = \sin \theta$ and $y = \cos \theta$, which are shown here for your reference

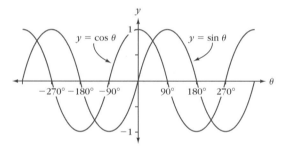

1. Consider the equation $\cos \theta = \sin (90° - \theta)$.

  a. Choose two values for θ, in different quadrants, and verify that the equation holds true for those values.

  b. Write an explanation of the equation in terms of the Ferris wheel model.

  c. Write an explanation of the equation in terms of the graphs of the sine and cosine functions.

Continued on next page

2. Consider the equation $\cos(-\theta) = \cos\theta$.

   a. Choose two values for θ, in different quadrants, and verify that the equation holds true for those values.

   b. Write an explanation of the equation using either the Ferris wheel model or the graph of the cosine function.

3. Consider the equation $\sin(-\theta) = -\sin\theta$.

   a. Choose two values for θ in different quadrants, and verify that the equation holds true for those values.

   b. Write an explanation of the equation using either the Ferris wheel model or the graph of the sine function.

DAY 15

The Cosine of a Sum

Mathematical Topics

- Justifying various formulas involving sine and cosine
- Seeing these relationships from graphs
- Discovering and justifying a formula for the cosine of the sum of two angles
- Using the sine-of-a-sum and the cosine-of-a-sum formulas to simplify the rotation formulas

Students develop the formula for the cosine of a sum.

Outline of the Day

In Class

1. Discuss *Homework 14: Oh, Say What You Can See*
   - A brief discussion is probably sufficient
   - Post the formulas for use today
2. Develop the cosine-of-a-sum formula
   - Lead students to derive the formula from previously derived formulas
   - Verify the formula in special cases
3. Develop a rotation formula using trigonometry
   - Lead students to derive formulas for coordinates of the rotated points in terms of the original coordinates and the angle of rotation

At Home

Homework 15: Comin' Round Again (and Again ...)

Special Materials Needed

- A transparency of the graphs of the sine and cosine functions (see Appendix C)

1. Discussion of *Homework 14: Oh, Say What You Can See*

For Question 1, have students share explanations of the equation $\cos \theta = \sin (90° - \theta)$ in terms of the Ferris wheel and the graph. Focus especially on why this equation holds for all angles (and not merely for acute angles).

Day 15

This is also a good opportunity to look at this equation in terms of the formal definition of sine and cosine. (Students may have done this in *High Dive*, but if they only verified it for some angles, we suggest that you have them look at this more carefully now.) You might bring out that when a point is rotated 90°, its *x*- and *y*-coordinates are essentially interchanged, up to sign. (One might view the angle $90° - \theta$ as the result of a reflection of the angle θ through the *x*-axis followed by a 90° counterclockwise rotation.)

It is also important that students feel fairly comfortable with the formulas $\cos(-\theta) = \cos\theta$ and $\sin(-\theta) = -\sin\theta$, because they will use these today to transform yesterday's formula for the sine of a sum into a formula for the cosine of a sum. Focus the discussion on how these relationships are illustrated in the graphs of the two functions.

We recommend that you post the trigonometric identities in this assignment, and use this occasion as an opportunity to review ideas from *High Dive* about the general definitions of sine and cosine. As noted on Day 14, you may want to have students keep individual lists of identities.

2. The Cosine of a Sum

Tell students that they are now ready to find the cosine of a sum of two angles, building on their work on *The Sine of a Sum*. Although you might want to let groups try to do this on their own, we suggest that this formula be developed primarily through a whole-class discussion. You can ask questions and offer hints to guide the class along, occasionally having groups work on pieces of the task.

To start, be sure that all of these formulas are readily available:

- $\sin(A + B) = \sin A \cos B + \cos A \sin B$
- $\cos\theta = \sin(90° - \theta)$
- $\sin\theta = \cos(90° - \theta)$
- $\cos(-\theta) = \cos\theta$
- $\sin(-\theta) = -\sin\theta$

Tell students that they will be putting these five formulas together to create an expression for $\cos(A + B)$. You might preface this work by asking if anyone has questions about any of these formulas.

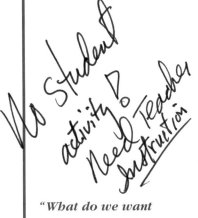

"What do we want on the left side of the equal sign?"

The first task is for students to relate the task of developing a cosine-of-a-sum formula to what they already know about the sine of a sum. Toward this end, ask students what the expression is for which they are trying to find a formula. Put another way, ask what they should put on the left side of the equal sign. The goal here is simply to bring out that they are trying to find an expression for $\cos(A + B)$ in terms of the individual sines and cosines of A and B.

114 As the Cube Turns

"Which of the five formulas could you use to get an equivalent expression for cos (A + B)?"

Next, ask students which of the five formulas they can use to get an equivalent expression for cos (A + B). As a hint, ask how they might rewrite this expression in terms of sine (for which they have a "sum" formula). This should lead to the idea of using the equation cos θ = sin (90° − θ), with A + B in place of θ. This should lead students to the equation

$$\cos(A+B) = \sin[90° - (A+B)]$$

"How could you write the right side as the sine of the sum?"

Tell students that in order to use the sine-of-a-sum formula, they need to write the right side of this equation as the sine of a sum. That is, they need to rewrite the expression 90° − (A + B) as a sum. Ask students to think about this for a bit. As a further hint, suggest that they write 90° − (A + B) simply as 90° − A − B, so the chain of development should now look like this:

$$\cos(A+B) = \sin[90° - (A+B)]$$
$$= \sin(90° - A - B)$$

If they don't come up with the idea of writing 90° − A − B as (90° − A) + (−B), then simply point out that these two expressions are equal. The chain of development should now look like this:

$$\cos(A+B) = \sin[90° - (A+B)]$$
$$= \sin(90° - A - B)$$
$$= \sin[(90° - A) + (-B)]$$

"What do you get if you apply the sine-of-the-sum formula to the right side?"

Next, ask groups what happens when they apply the sine-of-a-sum formula to the right side. This should lead them to develop the last line in this sequence:

$$\cos(A+B) = \sin[90° - (A+B)]$$
$$= \sin(90° - A - B)$$
$$= \sin[(90° - A) + (-B)]$$
$$= \sin(90° - A)\cos(-B) + \cos(90° - A)\sin(-B)$$

"Can you apply the other formulas to simplify the right side?"

Next, ask students to simplify this expression by using the last four items in the list of five formulas at the beginning of this discussion. Students should come up with these equations:

- $\sin(90° - A) = \cos A$
- $\cos(-B) = \cos B$
- $\cos(90° - A) = \sin A$
- $\sin(-B) = -\sin B$

Substituting the right sides of these equations into the last line of the chain of development so far, the class should end up with this sequence:

$$\cos(A + B) = \sin[90° - (A + B)]$$
$$= \sin(90° - A - B)$$
$$= \sin[(90° - A) + (-B)]$$
$$= \sin(90° - A)\cos(-B) + \cos(90° - A)\sin(-B)$$
$$= \cos A \cos B - \sin A \sin B$$

Put the beginning and end of this chain together as "Cosine of a Sum," and post this formula along with the sine-of-a-sum formula.

$$\cos(A + B) = \cos A \cos B - \sin A \sin B$$

- **Why is this formula correct?**

"How do you know this formula holds true for all angles?"

Ask students how they know that this formula holds true for all angles. If need be, point out that they derived this formula from formulas they had previously proved (at least for acute angles). Thus, their derivation of this formula is a proof of it, subject to some restrictions.

You can clarify that they only proved the sine-of-a-sum formula for acute angles (unless they have done the supplemental problems *Moving to the Second Quadrant, Adding 180°,* and *Sums for All Quadrants*). So they have only proved this cosine formula for some angles. But assure them that it is true for all angles, and that if time and patience permitted, they could prove it for all angles. The other formulas involved in the proof have been proved for all angles.

- **Verifying special cases**

As with the sine-of-a-sum formula, you can have students check some special cases. You can consider $B = 0°, 90°,$ and $180°$. Help students see (using the specific values of sine and cosine for these angles) that in these cases, the cosine-of-a-sum formula simplifies to these equations:

- $\cos A = \cos A$
- $\cos(A + 90°) = -\sin A$
- $\cos(A + 180°) = -\cos A$

The first is, of course, true for all values of A. You can suggest that students use the graphs of the sine and cosine function to develop an explanation for the second result. (They might also look at the supplemental problem *A Shift in Sine* from *High Dive,* which involves a slight variation on this identity.)

The third of these identities is considered in the supplemental problem *Adding 180°*. Students may be able to give a good explanation at this point in terms of the graph or the Ferris wheel model.

Day 15

3. Rotations Revisited

"Why are we interested in these formulas anyway?"

Ask students why they should be interested in these formulas. As a hint, ask where they stand so far in developing formulas for the result of rotation of a point. They should be able to point to these two formulas in terms of r and θ:

$$x' = r \cos(\theta + 10°)$$
$$y' = r \sin(\theta + 10°)$$

"What can you do with those expressions now?"

Have students work in groups. Ask them how they can apply their newfound formulas for sine-of-a-sum and cosine-of-a-sum to the formulas for x' and y'. They should get something like this:

$$x' = r(\cos\theta \cos 10° - \sin\theta \sin 10°)$$
$$y' = r(\sin\theta \cos 10° + \cos\theta \sin 10°)$$

"What do you get if you distribute the r through the expression?"

Next, ask them what happens when they distribute the r through the expression. This should give them these equations:

$$x' = r \cos\theta \cos 10° - r \sin\theta \sin 10°$$
$$y' = r \sin\theta \cos 10° + r \cos\theta \sin 10°$$

"Where have you seen $r \cos\theta$ before?"

Now ask them where they have seen $r \cos\theta$ before. They should be able to rewrite the formulas as

$$x' = x \cos 10° - y \sin 10°$$
$$y' = y \cos 10° + x \sin 10°$$

(using the relationships $y = r \sin\theta$ and $x = r \cos\theta$).

You may want to rewrite the second equation to put the term with x first:

$$y' = x \sin 10° + y \cos 10°$$

(This form will be easier to work with in terms of the matrix representation.)

With a bit of fanfare, point out that students have eliminated r and θ from the formula and written the new x- and y-coordinates directly in terms of the original x- and y-coordinates.

You may want to have students check this formula for the points $(3, 4)$ and $(2, 5)$ (assuming you saved the answers to *Homework 11: Goin' Round the Origin*) to see that it gives the right answers.

"Is there anything special about 10°?"

Before moving on, ask students if they think there is anything special here about 10°. Presumably, they will be comfortable with replacing this by a general value. To avoid confusion, you might want to introduce a Greek letter other than θ for the angle of rotation. We will use the letter ϕ, which is called *phi* (pronounced to rhyme with *pi*).

Interactive Mathematics Program

As the Cube Turns

Day 15

Post this general statement:

> **If a point (x, y) is rotated counterclockwise through an angle ϕ, then the coordinates (x', y') of its new location are given by the equations**
> $$x' = x \cos \phi - y \sin \phi$$
> $$y' = x \sin \phi + y \cos \phi$$

- *Verifying special cases*

"What do these formulas say if $\phi = 0°$? If $\phi = 90°$? If $\phi = 180°$?"

Once again, it's worthwhile to have students investigate whether these formulas make sense for some special cases. In particular, have groups look at what this result says for the cases $\phi = 0°, 90°,$ and $180°$.

Homework 15: Comin' Round Again (and Again ...)

> This assignment is basically an exercise in applying the new formulas.

"In my experience, I had only used trig identities to simplify trigonometric expressions and equations. So I was amazed that trig identities in As the Cube Turns *have other applications, including transformations and animation."*

IMP teacher Janice A. Bussey

Comin' Round Again (and Again ...)

To write a calculator program to rotate pictures, you need to be able to rotate points.

You have seen that when a point (x, y) is rotated counterclockwise through an angle ϕ, the coordinates (x', y') of its new location are given by the equations

$$x' = x \cos \phi - y \sin \phi$$
$$y' = x \sin \phi + y \cos \phi$$

To see how to use these formulas, make a small picture on a coordinate grid. Use only line segments in your picture to connect points. Use at least five points to get an interesting picture.

1. Write the x- and y-coordinates of the points in your picture.

2. Rotate your picture 60° around the origin by rotating each point. Don't guess about the new coordinates. Calculate them and show your work. Round off to the nearest tenth.

3. Based on your answers to Question 2, draw the rotated picture. (Be sure to keep track of which pairs of points you should connect by line segments.)

4. Rotate your new picture another 60° as you did previously, showing your work.

DAY 16

Rotation with Matrices

Students use matrices as the last piece to a nice formulation of rotations.

Mathematical Topics

- Reviewing matrix multiplication
- Using matrices to express rotations

Outline of the Day

In Class

1. Discuss *Homework 15: Comin' Round Again (and Again ...)*
   - If students have no questions, no discussion is needed
2. *More Memories of Matrices*
   - Students review multiplication of matrices and apply it to rotations
   - The activity will be discussed on Day 17

At Home

Homework 16: Taking Steps

1. Discussion of *Homework 15: Comin' Round Again (and Again ...)*

This should be a fairly routine assignment and a good indicator of whether students are comfortable with the work from yesterday. You need not take time to go over this unless students have questions. However, ask students to save their results. If time allows, they will compare them with an approach to rotations using matrices on Day 17. (See the subsection "Applying matrix rotations to specific examples" on Day 17.)

Interactive Mathematics Program   As the Cube Turns   121

Day 16

2. More Memories of Matrices

In this activity, students will review multiplication of matrices and then use matrix multiplication for rotation of points. If students get stuck when they get to Question 5, you may want to tell them explicitly to look for a 2×2 matrix M such that

$$[x \ y] \cdot M = [x \cos \phi - y \sin \phi \quad x \sin \phi + y \cos \phi]$$

Discussion of this activity is scheduled for tomorrow.

Homework 16: Taking Steps

> This assignment is a programming sidelight that students may find useful in their final projects.

More Memories of Matrices

In *Homework 9: Memories of Matrices*, you reviewed the idea of addition of matrices and saw how to apply that idea to the geometric translation of points.

Guess what? It's time to review some more about matrices and then to use them for rotations. Once again, the first question is based on an assignment from *Meadows or Malls?*

1. (From *Meadows or Malls?*) Linda Sue has a transport plane and delivers goods for two customers. One is Charley's Chicken Feed and the other is Careful Calculators. Here are some important facts about these two customers.

    - Charley's Chicken Feed packages its product in containers that weigh 40 pounds and are 2 cubic feet in volume.

    - Careful Calculators packages its materials in cartons that weigh 50 pounds and are 3 cubic feet in volume.

    a. Organize this information into a matrix, using the first row for Charley's Chicken Feed and the second row for Careful Calculators. Call this matrix *A*.

    b. Suppose that on Monday, Linda Sue transports 500 containers of chicken feed and 200 cartons of calculators. Put those facts into a row matrix (that is, a matrix with one row). Call this matrix *B*.

Continued on next page

c. Use the information in your two matrices to find out the total weight carried and the total volume used. Put those two answers into a row matrix. Call this matrix C.

Congratulations! If you completed Question 1c, you have multiplied two matrices. The matrix C is the product $B \cdot A$.

2. Based on your work in Question 1, write a description of how to multiply matrices. Your description should include a statement about when it is *possible* to multiply two matrices.

3. In Question 1, you found the matrix product $B \cdot A$. Now try to find the matrix product $A \cdot B$. Did you get the same result as in Question 1? Explain.

4. Suppose that on Tuesday, Linda Sue transports 400 containers of chicken feed and 300 cartons of calculators, and on Wednesday, she transports 250 containers of chicken feed and 350 cartons of calculators.

   a. Make a matrix showing Linda Sue's transportation record for all three days, using a different row for each day. Call this matrix D.

   b. Describe how you would use matrix multiplication to get a matrix showing the weight and volume carried each day.

5. Use the idea of matrix multiplication to express rotations. That is, find a way to use matrices to get from the ordered pair (x, y) to the ordered pair $(x \cos \phi - y \sin \phi, x \sin \phi + y \cos \phi)$.

Taking Steps

You've seen how the For/End instructions can be used to create a programming loop in which the loop variable increases by 1 each time the program goes through the loop.

For example, consider this plain-language program.

> For A from 4 to 8
> • Show A on the screen
> End the A loop

This program will print the numbers 4, 5, 6, 7, and 8 on the calculator screen.

Continued on next page

It's possible to make the loop variable increase by something other than 1 each time. For example, you can probably use an instruction on your calculator that looks something like this.

> For C from 2 to 11, step 3

This instruction will start the variable *C* at 2 and then increase it by 3 each time the program comes back to this instruction. The number 3 is called the *increment* or *step value*.

1. What do you think will appear on the screen when this program is run?

PROGRAM: STEP3

For C from 2 to 11, step 3
- Show C on the screen
End the C loop

2. Write a plain-language program that will show the calculator counting by 5's from 30 to 50.

3. What do you think would happen in each of these programs?

a. **PROGRAM: MISSTEP**

For W from 2 to 10, step 3
- Show W on the screen
End the W loop

b. **PROGRAM: FRACSTEP**

For S from 3 to 6, step 0.7
- Show S on the screen
End the S loop

4. How might you get the calculator to count backwards? For example, write a plain-language program (using a For/End loop) that you think would display the values 5, 4, 3, 2, and 1, in that order.

DAY 17

Students sum up ideas about rotations in two dimensions.

Rotation with Matrices, Continued

Mathematical Topics

• Analyzing variations in the use of the For/End instructions
• Reviewing matrix multiplication
• Using matrices to express rotations

Outline of the Day

In Class

1. Discuss *Homework 16: Taking Steps*
2. Discuss *More Memories of Matrices* (from Day 16)
   • Have students share ideas on how to remember the procedure for multiplying matrices
   • Develop and post a general formula for the rotation matrix
   • Have students apply rotation matrices to their work from *Homework 15: Comin' Round Again (and Again ...)*
3. Introduce *Homework 17: How Did We Get Here?*
   • Brainstorm terms students have learned so far in the unit

At Home

Homework 17: How Did We Get Here?

1. Discussion of *Homework 16: Taking Steps*

Ask two club card students to provide answers to Questions 1 and 2. For Question 1, students will probably agree that the calculator will print the numbers 2, 5, 8, and 11, and then stop. For Question 2, they are likely to mimic the program in Question 1, using instructions like these:

> For T from 30 to 50, step 5
> • Show T on the screen
> End the T loop

On Question 3, students will have had to guess a bit, because there is more than one reasonable way that a calculator might handle variations like these.

In effect, most calculators increase the loop variable each time the program comes back to the beginning of the loop (to the "For" instruction). If this increase makes the loop variable greater than the final value (which is 10 in the case of Question 3a), then the calculator considers the loop finished and goes directly to the instruction that comes after the end of the loop, that is, after the "End" instruction. (Thus, after a loop has been completed, the loop variable is inevitably larger than the final value.)

In other words, the result of the program in Question 3a is that the calculator displays the values 2, 5, and 8 only. Similarly, on Question 3b, the calculator displays 3, 3.7, 4.4, 5.1, and 5.8. (Have students check to be sure that this is how their calculators operate.)

Students' work on Question 4 will also have been speculative. If nobody came up with the idea of using a negative increment, you can leave this as an open question for students to investigate in their spare time. (Many calculators do allow the use of a negative increment, with the initial value larger than the final value.)

2. Discussion of *More Memories of Matrices*

- **Question 1**

    Let a spade card student present all of Question 1. The directions are fairly explicit about how to arrange the information, so students should come up with these matrices:

    $$A = \begin{bmatrix} 40 & 2 \\ 50 & 3 \end{bmatrix}$$

    $$B = \begin{bmatrix} 500 & 200 \end{bmatrix}$$

    $$C = \begin{bmatrix} 30{,}000 & 1{,}600 \end{bmatrix}$$

    The matrix multiplication involved in Question 1c is this equation:

    $$\begin{bmatrix} 500 & 200 \end{bmatrix} \cdot \begin{bmatrix} 40 & 2 \\ 50 & 3 \end{bmatrix} = \begin{bmatrix} 30{,}000 & 1{,}600 \end{bmatrix}$$

Day 17

"How exactly did you get 30,000 and 1,600?"

Ask students to explain which numbers are multiplied and what gets added to get 30,000 and 1,600. It may be helpful to explicitly write out equations like these:

$$500 \cdot 40 + 200 \cdot 50 = 30,000$$

$$500 \cdot 2 + 200 \cdot 3 = 1,600$$

- **Question 2**

    Ask two or three other spade card students to give their descriptions of how to multiply matrices.

"What mental images do you use to remember how to multiply matrices?"

You can also ask if students have any mental images that help them remember how to multiply matrices.

> For instance, some students may recall this idea, introduced in *Meadows or Malls?*: Think of the entries in a matrix as members of a marching band with numbers on their backs. To multiply the matrices, a row from the first matrix marches over into the second matrix, turns right at one of the columns, and moves into place to stand alongside the entries in that column.

- **Question 3**

    The matrix multiplication here is

    $$\begin{bmatrix} 40 & 2 \\ 50 & 3 \end{bmatrix} \cdot [500 \quad 200]$$

    which does not make sense. Review the fact that not every pair of matrices can be multiplied, and go over the conditions on the dimensions that make matrix multiplication possible. (The number of columns in the first matrix must be equal to the number of rows in the second matrix.)

"What does this show about matrix multiplication?"

Be sure to bring out explicitly that this shows that matrix multiplication is not commutative. (The term *commutative* was used in *Meadows or Malls?* but may need reviewing here.)

- **Question 4**

    You can have another spade card student present Question 4. The matrix D should look like this:

    $$D = \begin{bmatrix} 500 & 200 \\ 400 & 300 \\ 250 & 350 \end{bmatrix}$$

Day 17

The presenting student should be able to explain that the matrix of weights and volumes can be found as the product $D \cdot A$:

$$\begin{bmatrix} 500 & 200 \\ 400 & 300 \\ 250 & 350 \end{bmatrix} \begin{bmatrix} 40 & 2 \\ 50 & 3 \end{bmatrix} = \begin{bmatrix} 30{,}000 & 1{,}600 \\ 31{,}000 & 1{,}700 \\ 27{,}500 & 1{,}550 \end{bmatrix}$$

Use your judgment about whether you need an explicit explanation here as to which numbers were multiplied together.

- **Question 5**

   Finally, move on to the question of how to use matrices to represent the process of rotation of points. If many groups were stuck at this point yesterday, you may want to let them return to it now in their groups, with perhaps a clearer idea about matrix multiplication in general than they had yesterday.

   If a hint is needed, you might discuss the form of the equation. Help students to see that they are looking for a matrix M for which the product $[x\ y]M$ comes out to $[x \cos \phi - y \sin \phi\ \ x \sin \phi + y \cos \phi]$. They should see that they want

   $$M = \begin{bmatrix} \cos \phi & \sin \phi \\ -\sin \phi & \cos \phi \end{bmatrix}$$

   In other words, they should see that they can find the new coordinates using this matrix multiplication:

   $$[x\ y] \cdot \begin{bmatrix} \cos \phi & \sin \phi \\ -\sin \phi & \cos \phi \end{bmatrix} = [x \cos \phi - y \sin \phi\ \ x \sin \phi + y \cos \phi]$$

   Introduce the term **rotation matrix** for the matrix $\begin{bmatrix} \cos \phi & \sin \phi \\ -\sin \phi & \cos \phi \end{bmatrix}$.

- *Rotating many points*

   "How could you rotate several points together?"

   Finally, as a follow-up, ask students how this idea could be used to rotate many points. You may want to remind them that in *Homework 9: Memories of Matrices,* they put the coordinates for many points in a two-column matrix, with a separate row for each point.

   Here are further hints for the task of using matrix multiplication to rotate many points. First, have students set up a matrix of initial points, either numerically or using variables, to make the question clear. For example, you might suggest that they start with three points, (x, y), (s, t), and (u, v), and make a 3 × 2 matrix from them, which would look like this:

   $$\begin{bmatrix} x & y \\ s & t \\ u & v \end{bmatrix}$$

Next, ask what the final positions of each of these points would be if they were all rotated through an angle ϕ. Students should be able to see that the desired matrix is

$$\begin{bmatrix} x \cos \phi - y \sin \phi & x \sin \phi + y \cos \phi \\ s \cos \phi - t \sin \phi & s \sin \phi + t \cos \phi \\ u \cos \phi - v \sin \phi & u \sin \phi + v \cos \phi \end{bmatrix}$$

Finally, they need to see that they can get this matrix for the rotated points by multiplying the 3 × 2 matrix of points on the right by M, just as they did for a single point.

This result deserves a lot of fanfare and should be posted.

To rotate a matrix of two-dimensional points counterclockwise around the origin by an angle ϕ, multiply that matrix on the right by the matrix

$$\begin{bmatrix} \cos \phi & \sin \phi \\ -\sin \phi & \cos \phi \end{bmatrix}$$

(Students should also have this idea in their notes for use on later assignments.)

- *Applying matrix rotations to specific examples*

    To reinforce the method and strengthen students' faith in it, have them test the process out with one or more specific examples. Let the class choose the endpoints for a segment in the first quadrant and draw the segment on a coordinate grid.

    For simplicity, you can ask students to predict where each point would end up if rotated counterclockwise around the origin through an angle of 90°. They should multiply by the appropriate matrix and find out.

    If time allows, have students work in groups using the matrix multiplication process to check the answers they got on *Homework 15: Comin' Round Again (and Again …)*.

3. Homework 17: How Did We Get Here?

Tonight's homework is a reflective assignment, intended to help students sort out what they have learned so far in the unit. We suggest that you take a few minutes to brainstorm a list of mathematical terms that students have learned so far in this unit.

Day 17

How Did We Get Here?

In this unit, you have worked with several different mathematical concepts. Before you go on, it might be a good idea for you to sort things out. This assignment gives you a chance to reflect on where you have been and think about where you are going.

1. Make a list of the important mathematical terms and formulas you've learned or used so far in the unit. Write a definition of each of the key terms.

2. Write out the purpose of this unit, as clearly as you can.

3. For each item you listed in Question 1, discuss how it fits into the purpose of this unit.

DAY 18

Swing That Line!

Students apply their rotation formulas to calculator graphics.

Mathematical Topics

- Reviewing ideas from the unit
- Using rotation matrices in a program

Outline of the Day

In Class

1. Discuss *Homework 17: How Did We Get Here?*
   - Let students share ideas
   - Review the development of the rotation formula
2. *Swing That Line!*
   - Review programs from *Move That Line!* and the generic animation outline
   - Students write an animation program to rotate a line segment around the origin
   - The activity will be completed and discussed on Day 19
3. Introduce *Homework 18: Doubles and Differences*
   - Make sure students take home the collection of trigonometry formulas they have developed

At Home

Homework 18: Doubles and Differences

Special Materials Needed

- A copy of the program to move a line using matrices (from Day 10)

Reminder: On Day 20, you will need a sheet of Plexiglas or some other firm, clear material, for a preview demonstration of the activity *Cube on a Screen*. For the activity itself (on Day 27), you will need a sheet for each pair of students. You need to arrange to have the material available if it is not in your school already. Teachers in some schools have taped transparencies to hall or classroom windows or used the plastic covers that come with calculators.

Day 18

1. Discussion of *Homework 17: How Did We Get Here?*

Have students confer in groups on the homework. Then ask the heart card member of each group to suggest things they have learned.

Next, have one or two diamond card members describe what they see as the purpose of the unit. Students from other groups can critique the statements until a formulation is reached that most of the class is happy with. Then have diamond card students discuss how each item in the list of things learned fits into the purpose of the unit.

In particular, you can use this discussion as an opportunity to focus on the development of the rotation formula. Ask students about this series of activities—specifically, how they related to one another and how each contributed to the development of the formula:

- *Homework 10: Cornering the Cabbage*
- *Homework 11: Goin' Round the Origin*
- *Homework 12: Double Trouble*
- *The Sine of a Sum* (Days 13-14)
- *Homework 13: A Broken Button*
- *Homework 14: Oh, Say What You Can See*
- The "cosine of a sum" discussion on Day 15

2. *Swing That Line!*

Note: If you did not complete the discussion yesterday of *More Memories of Matrices,* then do that now, and use that as a lead-in to the discussion here.

Show students the program they made using matrices for *Move That Line!* You should have an overhead transparency or chart paper copy of this from Day 10, and students should have saved a copy of their own program.

134 As the Cube Turns Interactive Mathematics Program

Here is the program as described on Day 10:

> Setup program
>
> Let A be the matrix $\begin{bmatrix} 3 & 2 \\ 5 & 1 \end{bmatrix}$
>
> Let B be the matrix $\begin{bmatrix} 3 & -4 \\ 3 & -4 \end{bmatrix}$
>
> For N from 1 to 6
> - Clear the screen
> - Draw a line from (a_{11}, a_{12}) to (a_{21}, a_{22})
> - Delay
> - Replace matrix A by the sum A + B
>
> End the N loop

Ask for volunteers to describe what is happening in this program. One thing they should see is that the endpoints of the segment are put into a matrix called *A* (at the beginning of the program) and that these endpoints are later translated by the process of adding a special *translation matrix* to *A* (in the "Replace" line of the loop). This sum then becomes the new value of *A* and contains the endpoints of the translated segment.

You may also want to have students review the "generic" animation program outline they wrote on Day 9. The outline given there looked like this:

> Setup program
>
> Set the initial coordinates
>
> Start the loop
> - Clear the screen
> - Draw the next figure
> - Delay
> - Change the coordinates
>
> End the loop

Tell students their next activity is to write a program with a similar structure to *rotate* a line segment. As before, they will store the endpoints in a matrix.

"What else did you set (besides the initial coordinates)?"

"What else will you need to set in this program?"

Ask students what else they set at the beginning of the *Move That Line!* program besides the initial coordinates. Bring out that they also set the translation matrix, and ask what else they will need to set in this program. They should see that they will need to set the rotation matrix.

Day 18

Technical note: If students have saved their *Move That Line!* programs on their graphing calculators, they could begin the new program by changing the lines of that program. Most graphing calculators allow users to copy lines of an old program into a new program. You can show students how to do this on the calculator overhead projector. This will save them time and avoid tedious work.

We recommend that you have students work in pairs on *Swing That Line!* They will complete and discuss this activity tomorrow.

3. Homework 18: Doubles and Differences

In this homework assignment, students will create formulas for the sine and cosine of the difference of two angles and for the sine and cosine of twice an angle.

We expect that students will create these formulas by using formulas already developed, so they will need to have those formulas with them at home. Make sure they have these formulas in their notes and that they take these notes home tonight.

Use your judgment about whether simply to instruct students to take all their formulas home or to specify which formulas they will need. Here are the formulas that are required in the assignment.

- $\sin(A + B) = \sin A \cos B + \cos A \sin B$
- $\cos(A + B) = \cos A \cos B - \sin A \sin B$
- $\cos(-\theta) = \cos \theta$
- $\sin(-\theta) = -\sin \theta$

Swing That Line!

In this activity, your task is to write a program for the calculator that will take a line segment and rotate it counterclockwise around the origin a certain number of degrees. Your program should repeat this rotation several times, allowing you to see each segment briefly before erasing it and showing the next one.

1. Make a careful sketch on graph paper of the lines you want your calculator to draw.

2. Write a plain-language program to create the animation.

3. Turn your program from Question 2 into programming code for your calculator.

4. Enter and run your program from Question 3.

5. Once your program is successful, modify it to work with a more complicated picture.

Write down your completed programs on paper once you are satisfied with them on the calculator.

Suggestion: Use what you can of the *Move That Line!* program, simply making the necessary changes.

Doubles and Differences

You have developed formulas for the sine and cosine of the sum of two angles, writing sin $(A + B)$ and cos $(A + B)$ in terms of sin A, cos A, sin B, and cos B. In this assignment, you'll develop some variations on those formulas.

1. Develop a formula for the sine of the difference between two angles. That is, find a formula for sin $(A - B)$ in terms of sin A, cos A, sin B, and cos B. (*Hint:* Think of sin $(A - B)$ as sin $[A + (-B)]$.)

2. Check the formula you got in Question 1 by choosing several pairs of values for A and B and seeing if your formula works.

3. Find a formula for cos $(A - B)$ in terms of sin A, cos A, sin B, and cos B. Then check your formula by substituting pairs of values for A and B.

4. In *Homework 12: Double Trouble*, you got a formula for sin 80° in terms of sin 40° and cos 40° using ideas about area. Now, use the formula for sin $(A + B)$ to develop a formula for sin $2A$, in terms of sin A and cos A. Then check your formula by substituting values for A.

5. Use the formula for cos $(A + B)$ to develop a formula for cos $2A$. Then check your formula by substituting values for A.

DAY 19

Students complete and discuss their programs to swing a line.

Swing That Line Some More!

Mathematical Topics

• Developing trigonometric formulas
• Working with rotations using matrices in a program

Outline of the Day

In Class

1. Select presenters for tomorrow's discussion of *POW 4: A Wider Windshield Wiper, Please*
2. Discuss *Homework 18: Doubles and Differences*
3. Complete and discuss *Swing That Line!*
   • Have presentations when pairs have completed Question 3
   • Check off the second part of item 3 of the unit outline

At Home

Homework 19: *What's Going On Here?*

1. POW Presentation Preparation

Presentations of *POW 4: A Wider Windshield Wiper, Please* are scheduled for tomorrow. Choose three students to make POW presentations, and give them overhead transparencies and pens to take home to use for preparing presentations.

2. Discussion of *Homework 18: Doubles and Differences*

Ask students to come to consensus in their groups. Meanwhile, you can pass out transparencies for groups to prepare presentations. (You might omit having students show their work verifying the formulas.)

Interactive Mathematics Program

Then ask the club card members to make presentations. Here are the formulas:

- $\sin(A - B) = \sin A \cos B - \cos A \sin B$
- $\cos(A - B) = \cos A \cos B + \sin A \sin B$
- $\sin 2A = 2 \sin A \cos A$
- $\cos 2A = \cos^2 A - \sin^2 A$

3. Conclusion and Discussion of *Swing That Line!*

Wait until each pair has finished Question 3, if possible. You may want to have students show their programs on the calculator overhead projector, so they can actually run the programs as well as share the written code.

If students were working from the plain-language program for *Move That Line!* as shown in the discussion on Day 18, they might get a program like this, in which that same segment is rotated 10 times, through an angle of 15° each time:

Setup program

Let A be the matrix $\begin{bmatrix} 3 & 2 \\ 5 & 1 \end{bmatrix}$

Let B be the matrix $\begin{bmatrix} \cos 15 & \sin 15 \\ -\sin 15 & \cos 15 \end{bmatrix}$

For J from 1 to 11
- Clear the screen
- Draw a line from (a_{11}, a_{12}) to (a_{21}, a_{22})
- Delay
- Replace matrix A by the product A · B

End the J loop

Comment: The variable J goes from 1 through 11 here. The case $J = 1$ represents the initial drawing of the segment, and the cases from $J = 2$ through $J = 11$ represent the ten rotations.

It is time for another toot of your trumpet. Students can now check off the second part of item 3—"Rotate the object"—from their unit plan from Day 1 (as emended on Day 11).

Homework 19: What's Going On Here?

This assignment uses ideas that students have been working with, but applies them to a multipoint picture instead of to just a line segment.

What's Going On Here?

Once again, your task is to figure out what a certain plain-language program does. You should show on graph paper what would appear on the calculator screen after someone executes the program MYSTERY.

Program: MYSTERY

Setup program

Clear the screen

Let A be the matrix $\begin{bmatrix} 2 & 2 \\ 5 & 2 \\ 2 & 6 \\ 5 & 6 \\ 3.5 & 9 \end{bmatrix}$

Let B be the matrix $\begin{bmatrix} \cos 30 & \sin 30 \\ -\sin 30 & \cos 30 \end{bmatrix}$

For C from 1 to 12
- Draw a line from (a_{11}, a_{12}) to (a_{21}, a_{22})
- Draw a line from (a_{21}, a_{22}) to (a_{41}, a_{42})
- Draw a line from (a_{41}, a_{42}) to (a_{51}, a_{52})
- Draw a line from (a_{11}, a_{12}) to (a_{31}, a_{32})
- Draw a line from (a_{31}, a_{32}) to (a_{51}, a_{52})
- Replace matrix A by the product $A \cdot B$

End the C loop

Day 20

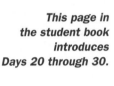

DAYS 20-30

Projecting Pictures

This page in the student book introduces Days 20 through 30.

Cassie Burniston and Sam Ellison are enmeshed in how to create a two-dimensional drawing of a three-dimensional cube.

You are now ready to move on to three dimensions. You will need to review some things you learned in *Meadows or Malls?* about three-dimensional graphs. You will also need to adapt some of the two-dimensional geometry you learned in the Year 3 unit *Orchard Hideout* to three dimensions.

Interactive Mathematics Program   As the Cube Turns   159

DAY 20

POW 4 Presentations

Students begin to think about projections and do presentations of POW 4.

Mathematical Topics

• Understanding calculator programs that use matrix transformations
• Introducing the idea of projection onto a plane
• Using a line of sight to understand projections

Outline of the Day

In Class

1. Discuss *Homework 19: What's Going On Here?*
   • Have students go through the program step by step
   • Have students create code for the program and then enter and run it

2. Introduce the idea of projection onto a plane
   • Discuss the need to draw three-dimensional objects on a two-dimensional space
   • Demonstrate how to use a line of sight to make a projection

3. Presentations of *POW 4: A Wider Windshield Wiper, Please*

4. Introduce *POW 5: An Animated POW*
   • Specify options, limitations, and grading criteria
   • Present students with a timetable for stages of the project

At Home

Homework 20: *"A Snack in the Middle" Revisited*
POW 5: An Animated POW (due Day 35)

Special Materials Needed

• A sheet of Plexiglas or other firm, clear material
• A cube (at least 2 inches on an edge)

Discuss With Your Colleagues

Is the Time Spent on the Project Worthwhile?

Students will be spending considerable time preparing and presenting work on *POW 5: An Animated POW*. What might your students get out of this project? Do the results justify the time spent?

Interactive Mathematics Program — As the Cube Turns 145

Day 20

1. Discussion of *Homework 19: What's Going On Here?*

Have various students go step by step through the program, describing what happens and perhaps going through the loop two or three times. We suggest that you have the class set up a table to keep track of the changing values for the matrix *A* and the loop variable *C* (*B* doesn't change), and plot the successive drawings on coordinate axes.

Then have the class create code for the program, enter it, and run it. The result should look something like this:

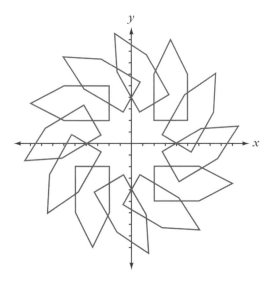

That is, the program draws an initial "house" figure, as shown in the next diagram. Then it rotates that figure around the origin in 30° increments, creating 12 separate versions of the picture.

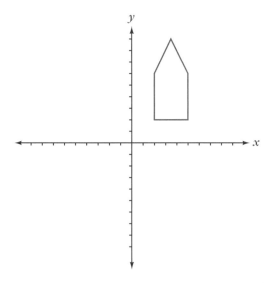

Be sure students see how the house is turning around the origin. For example, the distance from each corner of the house to the origin remains

146 *As the Cube Turns*

constant throughout the process, and the corner closest to the origin stays closest.

"How could the program be modified to show only one picture at a time?"

You can ask students how the program could be modified to show only one picture at a time, to create more of an illusion of motion. They should see that they could add a delay loop and an instruction to clear the screen within the existing For/End loop. You can suggest that they work on this variation in their spare time.

"What causes the figure to make one complete circle with its motion?"

You might also ask what causes the figure to make one complete circle with its motion, or how the figure would need to be altered if the angle for each rotation were some value other than 30°.

2. Introduction to Projections

It's time for another look, a brief one for now, at the general plan of the unit. You can start with the generic animation outline developed on Day 9, which should look something like this:

    Setup program

    Set the initial coordinates

    Start the loop
- Clear the screen
- Draw the next figure
- Delay
- Change the coordinates

    End the loop

"What would you need to do to write the program for the turning cube?"

Ask students what they might need to do to carry out this plan for the unit problem of the turning cube. They may start by looking at the step of the outline in which they set the initial coordinates. In other words, they should see that they need to decide where the cube is located at the beginning of the program.

You can tell them that they will get to this in a while. Then turn their attention to the next nontrivial step, namely, drawing the next figure.

"How do you draw a three-dimensional object on a two-dimensional screen?"

Ask why this might be difficult. You want to bring out that the cube itself is a three-dimensional figure, but that their drawing on the calculator screen will have to be a two-dimensional figure. Tell students that this process is called **projecting the figure** and that such a two-dimensional drawing is called a **projection.** Bring out that this task is item 4 of the unit plan: "Create a two-dimensional drawing of a three-dimensional object."

Let students know that this is not an easy task and that they will spend the next week or so on activities to help them understand how to accomplish it.

Tell them that as they work on these activities, beginning with tonight's homework, they should be trying to figure out how these activities are related to the idea of projecting from three dimensions to two.

We recommend that you give the class a brief preview demonstration now of an activity that they will do on Day 27, called *Cube on a Screen*. The activity and this preview use a cube and a sheet of Plexiglas (or similar firm, clear material on which students can write). You will need three volunteers to carry out the demonstration.

Have one volunteer hold the sheet of Plexiglas, have a second hold the cube behind the Plexiglas, and have the third stand in front of the Plexiglas. Then have the first volunteer point to one of the vertices of the cube, and ask the third volunteer to determine about where on the Plexiglas he or she would draw that vertex, based on how it appears as viewed through the Plexiglas.

The goal here is to suggest that the location on the Plexiglas where the point should be drawn is determined by the line from the viewer's eye to the vertex. The class's challenge over the next week or so will be to figure out how to express this location on the Plexiglas in terms of coordinates of the vertex and the viewer's eye. However, students need to do considerable work on related ideas before they address this issue directly.

3. Presentations of *POW 4: A Wider Windshield Wiper, Please*

Ask the three selected students to make their presentations. The two wipers don't clean the same area, and it will be interesting to see whether students can find a wiper that covers more area than either of the models described.

A flawed explanation

You may find it interesting to examine the following argument, which appears to show that the two wipers clean the same area.

The standard wiper cleans one-fourth of the area between the two concentric circles with radii of 18 inches and 6 inches. So that wiper cleans an area of 72π square inches.

The second wiper is also a blade with length 12 inches. As it moves from left to right, its midpoint moves through a path that is one-fourth of the circumference of a circle with radius 12 inches. (This is the path of the tip of the arm.) This path has length 6π inches. So the blade is sweeping a 12-inch vertical strip across a path of length 6π inches, which gives an area of 72π.

It turns out that the second part of this argument is incorrect. What exactly is wrong with this argument, besides the fact that you get the wrong answer? (*Hint:* Think about comparing the area of a rectangle with the area of a parallelogram. The area covered by the second wiper is actually $144\sqrt{2}$ square inches.)

4. Introducing *POW 5: An Animated POW*

Remind students that at the end of the unit, they will write an animation program. *POW 5: An Animated POW* describes this animation project. Students will work in pairs, but each student needs to turn in an overall outline for the program, a more detailed plain-language version of the program, and the actual code for the program. You should take time now to clarify again what levels of detail you expect for each of these components.

- *Options for the POW*

    The POW asks students to write a calculator program. However, you might allow them other options such as writing a computer program or doing some animation using cutouts and a video camera, leaving the choice of medium to them so as not to stifle their creativity.

    Also, as written, the POW does not specify the mathematical expectations for the project. Feedback from teachers suggests that students may get a great deal out of the assignment even if they do not make significant use of the new mathematical ideas of the unit. If you want to have students apply some of the mathematical ideas of the unit, you might give them a specific guideline, such as requiring them to use a matrix in some way or asking them to use at least one mathematical idea and report on it in their presentations.

    It is probably a good idea to tell students what your grading criteria will be for the POW. Here are some possible items to include.

    - Creativity
    - Aesthetics of the visual displays of their programs
    - Elegance of their programs
    - Use of mathematics

- *Schedule for the POW*

    We suggest that you give students a schedule for completing the various stages of the POW. The schedule listed here is incorporated into this teacher guide.

    - Day 20 (today): Introduce the POW.
    - Day 22: Students hand in their partner selections.
    - Day 24: Partners hand in their descriptions of what their programs will do.
    - Day 29: Partners hand in a written outline of their program (*Homework 28: An Animated Outline* asks students to finish up their outlines.)

Day 20

- Day 34: Day 34 and the homework that night are set aside for students to complete their POWs.
- Days 35–36: Partners make presentations to the class and hand in their written programs.

• *Use of graphing calculators*

You should discuss what arrangements can be made for students to use the calculators outside of class (during lunch, after school, at home overnight, and so on).

Homework 20: "A Snack in the Middle" Revisited

> This assignment builds on the situation in *Homework 7: A Snack in the Middle* from the Year 3 unit *Orchard Hideout*, in which the focus was on developing and at least partially proving the midpoint formula. In tonight's assignment, the desired point is still on a given line segment, but not necessarily halfway between the endpoints. The focus of the assignment is on how to modify the midpoint formula as the "fraction of the way" changes. This work will be continued in tomorrow's activity, *Fractional Snacks*.
>
> This assignment is the first step toward item 4 of the unit plan—"Create a two-dimensional drawing of a three-dimensional object"—although the connection will probably not be immediately clear to students.

If you have students who are not familiar with the Year 3 unit *Orchard Hideout*, you should take a minute to explain the setup of the orchard, with trees planted at lattice points in a coordinate system. (You do not need to explain the entire unit problem.)

You may want to suggest that students use graph paper for this assignment.

"One of my favorite activities of As the Cube Turns *is when students draw the projection of the cube onto a sheet of Plexiglass. They squint one eye shut, stick their tongues out between their teeth, and grimace studiously as they attempt to properly orient the cube on the plastic. The exercise is a nice use of physical manipulatives in a unit that is mostly cerebral."*

IMP teacher Dave Robathan

POW 5

An Animated POW

Your assignment in this POW is to create an interesting animation program for the graphing calculator. You will work with a partner.

You need to hand in a general outline of the program, a more detailed plain-language program, and the actual code for the program. Be sure to keep a written copy of your work at all times. (It turns out that even merely dropping the calculator can cause all of its programs to be erased!)

You and your partner will make a presentation of your animation to the class, lasting three to four minutes. You will show your animation on the overhead projector, and you will describe one interesting feature of how you did the programming.

Here are the different stages of your work on this POW:

- You give your teacher a statement of who your partner is.

- You and your partner hand in a description of what you want your program to do.

- You and your partner make a presentation to the class, lasting three to four minutes. Your presentation should include

    ✓ a demonstration of your program running on the overhead calculator

    ✓ a description of one interesting feature of your program

- You hand in the written outline, plain-language program, and code.

"A Snack in the Middle" Revisited

You may recall from *Orchard Hideout* that Madie and Clyde sometimes spent their afternoons pruning their trees. Each afternoon, they would choose two trees that seemed most in need of pruning.

Pruning made them hungry, so they would set up a snack between the two trees they were working on. They knew they would both want to have snacks at various times during the afternoon, so they agreed to set up their snack at the midpoint of the segment connecting the two trees.

1. If they were working on the trees at (24, 6) and (30, 14), where should they set up the snack? Explain your answer.

2. If they were working on the trees at (6, 2) and (11, 9), where should they set up the snack? Explain your answer.

One afternoon, Clyde was daydreaming of his delicious snack, and he slipped from his ladder and cracked his kneecap. Madie graciously offered to do the pruning herself, but Clyde would not hear of it.

Madie also offered to keep the snacks at whatever tree Clyde was working on, but he wouldn't agree to that, either. Clyde finally agreed to have the snacks placed one-third of the way from himself to Madie, instead of halfway.

3. The next time they went out, Clyde was working on the tree at (6, 9) and Madie was working on the tree at (18, 15). According to their new agreement, where should they put the snack? Explain your answer.

DAY 21

Orchard Snacks

Mathematical Topics

- Finding points a certain fraction of the way along a line segment
- Using similar triangles

Students generalize the midpoint formula for arbitrary fractions.

Outline of the Day

In Class

1. Remind students that names of POW partners are due tomorrow
2. Discuss *Homework 20: "A Snack in the Middle" Revisited*
   - For Question 1, focus on the proof that the midpoint formula works
   - Use a diagram to illustrate the similar triangles involved in Question 3
3. *Fractional Snacks*
   - Students generalize the midpoint formula for arbitrary fractions
4. Discuss *Fractional Snacks*
   - Post the general formula
   - Ask students how this work relates to the unit problem
   - Bring out that the midpoint formula is a special case of this new formula

At Home

Homework 21: *More Walking for Clyde*

Special Materials Needed

- Transparencies of the diagrams of the snack situations (see Appendix C)

1. POW Partner Reminder

Remind students that tomorrow they should hand in their selections of partners for work on *POW 5: An Animated POW.* You might have each pair turn in a piece of paper signed by both partners.

Interactive Mathematics Program   As the Cube Turns   153

Day 21

2. Discussion of *Homework 20*: *"A Snack in the Middle" Revisited*

Comment: The discussion of this and the next couple of assignments focuses on issues in the two-dimensional coordinate system in preparation for later work in three dimensions. It is very important that students feel comfortable with the ideas in this simpler setting before they move on to a situation that is often much harder for them to visualize.

In the discussion here, we focus on the proof in connection with Question 1 and then more on the computation on Questions 2 and 3.

- Question 1

"How did you determine your answer?"

Have a spade card student report on Question 1. Be sure to have the student explain how she or he found the numerical answer. Then ask for volunteers who did the computation differently. It will be helpful to see a variety of ways.

Here are three likely approaches to Question 1. You should try to coax at least some variety out of the class.

- Students may remember the midpoint formula, which would lead them to use the computation $\left(\frac{24+30}{2}, \frac{6+14}{2}\right)$.
- Students may do something similar on a more intuitive level, saying that 27 is halfway between 24 and 30 and that 10 is halfway between 6 and 14.
- Students may take half the difference in coordinates in each direction and add these on to the values at (24, 6). For example, they may say that (30, 14) is 6 units to the right of (24, 6), so the midpoint should be 3 units to the right of (24, 6). This makes the *x*-coordinate of the midpoint equal to 24 + 3, which is 27. Similarly, the *y*-coordinate of the midpoint is equal to $6 + \frac{1}{2} \cdot 8$, which is 10.

- *The x- and y-coordinates work independently*

Before going on to Questions 2 and 3, bring out that for all of these methods, students are working with the *x*-coordinates and *y*-coordinates separately. Ask if anyone can *prove* that any of these methods are correct. (*Note:* In *Orchard Hideout*, students proved that this formula gives a point that is equidistant from the two initial points, but they may not have proved that it gives the midpoint, that is, that the resulting point is on the segment connecting the initial points.)

154 *As the Cube Turns*

Interactive Mathematics Program

As a hint, if necessary, suggest that they use a diagram like the one shown here, where (?, ?) represents the midpoint. (A large version of this diagram is included in Appendix C.)

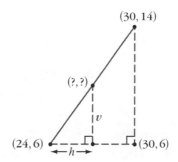

The small right triangle is similar to the large one, because they have a common angle at (24, 6) and each has a right angle. But the hypotenuse of the small triangle is, by assumption, half the length of the large hypotenuse. Therefore, other parts of the small triangle are half the corresponding parts of the large triangle.

Because the vertical leg of the large triangle has length 8 (= 14 − 6), the vertical leg of the small triangle (shown as v) must have length 4. Similarly, the horizontal leg of the small triangle (shown as h) has length 3. Thus, the midpoint is (27, 10). (*Note:* Putting the coordinates of the point at (30, 6) into the diagram will probably help many students find the lengths 8 and 6.)

You need not get bogged down in the details of this. The main point is for students to recognize that the computation rests on ideas about similar triangles. At various times in the work on item 4—"Create a two-dimensional drawing of a three-dimensional object"—you will probably want to ask about the geometric principle behind a computation. Students should be able to identify similarity as the guiding concept.

Save this diagram so you can refer to it when the discussions related to three dimensions occur.

- Question 2

    Unless students had considerable difficulty with Question 1, you can probably simply have another spade card student go over the computation for Question 2, again looking for different methods.

- Question 3

    Let another spade card student explain how he or she did Question 3. Most likely, the student will use something analogous to the third of the methods outlined for Question 1, saying that Madie is 12 units to the right of Clyde, so the snack should be 4 units to Clyde's right ($\frac{1}{3}$ of 12 units). Similarly, the snack should be 2 units "up" from Clyde ($\frac{1}{3}$ of 6 units).

Day 21

You should get students to draw a diagram and then identify similar triangles. The diagram might look like the one shown here. (This diagram also is included in Appendix C.)

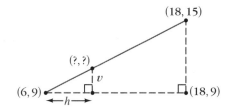

Here, we are told that the distance from (6, 9) to (?, ?) is $\frac{1}{3}$ of the distance from (6, 9) to (18, 15). Therefore, because of similar triangles, h is $\frac{1}{3}$ of 12 and v is $\frac{1}{3}$ of 6.

Note: Some students may say that the snack should be at (4, 2), using the values of h and v, and forgetting about the final step of adding the 4 units to Clyde's x-coordinate and the 2 units to his y-coordinate. Be sure to identify this step in the computation.

3. Fractional Snacks

You can tell students that in this activity, they will generalize the ideas from the homework. No other introduction is needed.

> *Note:* We use the letter r in Question 3 to suggest the word "ratio."

4. Discussion of *Fractional Snacks*

• Question 1

Let a heart card student explain how she or he did Question 1. The method will probably be quite similar to that used on Question 3 of the homework.

"How was this computation different from that in Question 3 of the homework?"

Ask how the computation in this problem differed from that in Question 3 of the homework. This can help students identify how they are using the fraction $\frac{1}{4}$ from the problem.

You may also want to use this problem to get students to focus clearly on the computation by which they get the two components of the distance from Clyde to Madie. For example, if the presenting student says that Madie is 16 units to the right of Clyde, ask how the student found that distance. Getting students to express this explicitly as $14 - (-2)$ will help them in formulating the general case.

156 *As the Cube Turns* Interactive Mathematics Program

Day 21

- **Question 2**

    There are two changes to watch out for as students discuss this problem.

    - In previous questions, students may have divided the appropriate distances by either 3 or 4, rather than multiply by $\frac{1}{3}$ or $\frac{1}{4}$. They may have some trouble with Question 4 because the fraction involved is not a *unit fraction* (that is, a fraction with numerator equal to 1). You may need to talk about multiplying by the appropriate fraction rather than dividing.

    - In the earlier problems, each of Madie's coordinates was larger than Clyde's corresponding coordinate. Students may have some trouble in Question 2 dealing with the signs of the differences between the coordinates.

    Both of these issues deserve attention, especially in preparation for the generalization. You can use the discussion of Question 2 to bring out, for instance, that one can always subtract Clyde's coordinates from Madie's, multiply by the fraction, and add the results to Clyde's coordinates. If Madie's coordinates are smaller than Clyde's, it simply means that the differences will be negative.

- **Question 3**

    We suggest that you ask for volunteers on Question 3. It is more difficult than Questions 1 and 2, so students may have worked on it at different levels.

    Some students may express the description verbally, along these lines:

    > "Find the distances from Clyde to Madie in each of the *x*- and *y*-directions, then take the right fraction of these distances, and then add these fractions to Clyde's position."

    Other students may formulate the process algebraically. For example, they may give both Clyde's and Madie's initial coordinates symbolically and then give a formula for the coordinates of the snack.

    Be sure to validate both approaches, but work toward getting the whole class to see the algebraic approach, building on the concrete examples. Help students focus on two specific elements:

    - The general approach of "original coordinates plus a fraction of the distance between them"

    - The role of similar triangles, including the fact that the same "fraction of the distance" is seen for each pair of sides of the similar triangles

Day 21

Students should develop a generalization something like this:

POST THIS!

If r is any fraction, then the point that is "r of the way" from (x_1, y_1) to (x_2, y_2) has these coordinates:

$$(x_1 + r(x_2 - x_1), y_1 + r(y_2 - y_1))$$

Post this general rule. We suggest that you also have students connect this formula back to their verbal description. Thus, $x_2 - x_1$ and $y_2 - y_1$ represent "finding the distances from Clyde to Madie in each of the x- and y-directions," multiplying by r represents "taking the right fraction of these distances," and the sums $x_1 + r(x_2 + x_1)$ and $y_1 + r(y_2 - y_1)$ represent "adding these fractions to Clyde's position."

• *The midpoint formula*

"What does this formula say about finding a midpoint?"

Regardless of whether students referred explicitly to the midpoint formula in the discussion of Question 1 of the homework, ask them what the general formula just developed tells them about the case of finding the midpoint.

One purpose of this question is to have them recognize that the midpoint is the special case $r = \frac{1}{2}$. The other purpose is to give them an opportunity to apply the general formula, substituting $\frac{1}{2}$ for r, and simplify the algebra. That is, they should see that the expression

$$\left(x_1 + \frac{1}{2}(x_2 - x_1), y_1 + \frac{1}{2}(y_2 - y_1)\right)$$

simplifies to

$$\left(\frac{x_1 + x_2}{2}, \frac{y_1 + y_2}{2}\right)$$

which is the usual form of the midpoint formula.

• *Snacks and turning cubes*

"How does this formula relate to the unit question?"

Tell students that they should be thinking about what the snack problems have to do with the problem of programming a turning cube. This does not need to be discussed now, although students may realize that this formula will help them compute the coordinates of a projected point.

Homework 21: More Walking for Clyde

> This assignment continues the general idea of last night's homework and today's activity.

158 *As the Cube Turns* Interactive Mathematics Program

Fractional Snacks

You saw in *Homework 20: "A Snack in the Middle" Revisited* that Clyde had injured his knee. Because he was having trouble getting around, he and Madie decided to place their afternoon snack closer to Clyde than to Madie.

This activity continues with the lives of our busy pruners on the following day.

1. Unfortunately, Clyde felt even worse the next day. So he and Madie decided that the snacks would be only one-fourth of the way from Clyde to Madie. If Clyde worked on the tree at $(-2, 7)$ and Madie worked on the tree at $(14, 19)$, where should they put the snack?

2. Then Clyde began to feel a little better. So they increased the fraction so that the snack would be two-fifths of the way from Clyde to Madie. This time Clyde's position was $(6, 13)$ and Madie's was $(9, 1)$. Where should they put the snack?

3. As Clyde's condition varied, he and Madie kept changing the fraction that they used in deciding where to place the snack. Describe how to compute the coordinates of the snack if the given "fraction of the distance" is r.

More Walking for Clyde

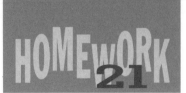

Clyde's knee is pretty well healed, but he needs to walk regularly to strengthen his leg muscles. Thus, Clyde and Madie have decided to place the snack so that Clyde actually has to walk past Madie to get to it.

For this assignment, assume that Clyde is at (10, 12) and that Madie is at (16, 2). The snack will still be on the straight line that connects their two positions.

1. Suppose Clyde and Madie decide to make the distance from Madie to the snack equal to half of her distance from Clyde. Where should they put the snack?

2. Suppose Clyde and Madie decide to make the distance from Madie to the snack equal to twice her distance from Clyde. Where should they put the snack?

3. In general, if the distance from Madie to the snack is t times the distance from Madie to Clyde, where should they put the snack?

DAY 22

Monorail Delivery

A new twist is given to the problem of finding a fractional distance from two points.

Mathematical Topics

• Continuing work finding points a certain fraction of the way along a line segment

Outline of the Day

In Class

1. Collect names of POW partners
2. Discuss *Homework 21: More Walking for Clyde*
3. *Monorail Delivery*
   • Have students read the problem as a class before beginning group work, and discuss how it differs from the previous snack problems
   • Students find the point where a given vertical line meets a given line segment
4. Discuss *Monorail Delivery*
   • Focus on how to use the known *x*-coordinate to find the desired ratio

At Home

Homework 22: Another Mystery

1. Collect POW 5 Partner Names

Have students turn in the names of their partners for work on *POW 5: An Animated POW.*

2. Discussion of *Homework 21: More Walking for Clyde*

The problems in last night's homework are quite similar to those in *Homework 20: "A Snack in the Middle" Revisited*. There are several approaches students might use, and you should elicit a variety of ideas.

Day 22

If students work with the numbers as given in the problems (that is, using $\frac{1}{2}$ in Question 1 and 2 in Question 2), they should realize that the *x*- and *y*-components need to be added to Madie's coordinates instead of to Clyde's.

For example, in Question 1, using a diagram like the one shown here, students can see that h is half of $16 - 10$ and that v is half of $2 - 12$. These values are then added to Madie's coordinates to show that the snack belongs at $(16 + 3, 2 + (-5))$, which is $(19, -3)$. (*Note:* Students may prefer to think of v as 5 rather than as -5 and to subtract 5 from 2 rather than add -5 to 2.)

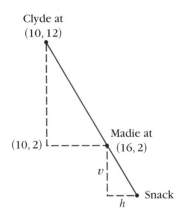

An alternate approach for Question 1 is to say that the snack's distance from Clyde is $1\frac{1}{2}$ times the distance from Clyde to Madie, and use the number $1\frac{1}{2}$ as the value for r in the formula developed yesterday.

> *Note:* Use your judgment about whether to develop a general formula for the "beyond Madie" situation. It will not be needed in this unit.

3. Monorail Delivery

The activity *Monorail Delivery* continues the dilemma of where Clyde and Madie should put their snack. The new twist here is that students need to use the information in the problem to find the fraction that corresponds to the number r in yesterday's formula. This fraction was given to students in the previous snack problems. (When students develop their programs to turn the cube, finding a similar fraction will be an important element of the process.)

We suggest that you have students read the problem (to themselves or taking turns reading aloud) and then discuss it briefly before they begin work.

"How is this problem different from the homework? What information do you have, and what don't you have?"

After students have finished reading the problem, ask the whole class how it differs from the homework. If necessary, ask what information they have and what they need to figure out. Help them as needed to see that in the previous snack problems, they knew what fraction of the way the snack was located on the line between Clyde and Madie.

162 *As the Cube Turns*

If a hint is needed, ask:
"How could you find the fraction needed?"

With that clarification, you can have groups begin work on the problem. If a further hint is needed, ask how they might find that fraction in the *Monorail Delivery* problems. If necessary, ask where else the fraction appeared in the previous problems (aside from its explicit mention). Students might remember that the fraction also shows up in a comparison of the vertical and horizontal sides of a small triangle to those of a similar large triangle.

In the problems in *Monorail Delivery,* students can find the ratio of horizontal sides directly because the problem says, in effect, that the *x*-coordinate of the snack is 12.

4. Discussion of *Monorail Delivery*

Let diamond card students from different groups present Questions 1 and 2. For Question 1, students might use a diagram like this one.

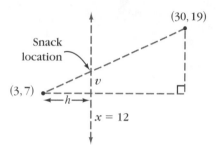

In this diagram, the distance h can be found directly as $12 - 3$, because the monorail is along the line $x = 12$. The key step is realizing that the fraction called r in the earlier formula is given by the ratio $\frac{12-3}{30-3}$ (which is $\frac{1}{3}$). This fraction can then be used to find v (which is one-third of $19 - 7$), and then the value of v can be added to 7 to get the *y*-coordinate of the snack's location. Thus, for Question 1, the snack should be dropped off at the point (12, 11). Question 2 is similar.

Note: Some students might want to get the equation of the line connecting the two points and use that to find the intersection with the monorail. If that comes up, acknowledge that it is a good idea in this case, but tell students that the method will not generalize well to the three-dimensional issues awaiting them in the unit problem.

After the class has discussed Questions 1 and 2, ask for a volunteer to present ideas on how to create a general formula (still using $x = 12$ as the equation of the monorail). As needed, help the class work from the specific examples to get that in general, the fraction can be expressed as $\frac{12-a}{c-a}$. If this fraction is labeled r, then the position of the snack can be expressed as

$$(12, b + r(d - b))$$

Day 22

- *Optional: A further generalization*

    In preparation for later work in this unit, you may want to have students generalize this expression further, using an arbitrary line $x = k$ to represent the monorail. In this case, r is the fraction $\frac{k-a}{c-a}$, and the position of the snack can be expressed as $(k, b + r(d - b))$.

Homework 22: Another Mystery

> This program is fairly easy to follow conceptually, but working out the details of exactly what gets drawn is not so easy. This is the first example students have seen that combines a rotation with a translation.

Monorail Delivery

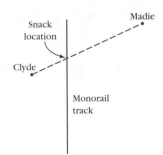

When Clyde recovered from his accident, his next project was to help Madie build a monorail in the orchard. (A monorail is a train that runs along a single track.)

Madie and Clyde planned to use the monorail to help them transport the apples they grew to a central location for shipping to market.

The first monorail they built ran along the line $x = 12$, and the train ran along this track in both directions. Clyde and Madie realized that they could also use the monorail to bring themselves fresh snacks when they were out in the orchard pruning trees. The monorail would drop off their snacks at the point where the line that joined their locations crossed the track. (They arranged to work on opposite sides of the track, and they had stopped worrying about who would be closer to the snack.)

1. If Clyde is working on the tree at (3, 7) and Madie is working on the tree at (30, 19), where should the monorail drop off the snack? Make a sketch of the situation and explain your answer.

2. If Clyde is working on the tree at (−5, 28) and Madie is working on the tree at (18, −14), where should the monorail drop off the snack? Make a sketch of the situation and explain your answer.

3. Generalize your work from Questions 1 and 2. If Clyde is working on the tree at (a, b) and Madie is working on the tree at (c, d), where should the monorail drop off the snack?

Another Mystery

This assignment shows another plain-language program for the calculator. Your job is to figure out what it does and then create programming code for it.

Note: The order of operations for matrix arithmetic is the same as the order of operations for number arithmetic.

PROGRAM: ANOTHER

Setup program

Clear the screen

Let A be the matrix $\begin{bmatrix} 1 & 1 \\ 4 & 1 \\ 2 & 4 \end{bmatrix}$

Let B be the matrix $\begin{bmatrix} 6 & -1 \\ 6 & -1 \\ 6 & -1 \end{bmatrix}$

Let C be the matrix $\begin{bmatrix} \cos 180 & \sin 180 \\ -\sin 180 & \cos 180 \end{bmatrix}$

For W from 1 to 4
- Draw a line from (a_{11}, a_{12}) to (a_{21}, a_{22})
- Draw a line from (a_{11}, a_{12}) to (a_{31}, a_{32})
- Draw a line from (a_{21}, a_{22}) to (a_{31}, a_{32})
- Replace A by the matrix $B + A \cdot C$

End the W loop

1. Make a careful drawing on graph paper of what the calculator screen will show after someone executes the program ANOTHER.

2. Create programming code for your calculator that will accomplish what this plain-language program describes.

DAY 23

Return to the Third Dimension

Students review the three-dimensional coordinate system.

Mathematical Topics

- Analyzing a program combining rotation and translation using matrices
- Reviewing three-dimensional graphing
- Visualizing objects in three dimensions

Outline of the Day

In Class

1. Form new random groups
2. Remind students that POW plans are due tomorrow
3. Discuss *Homework 22: Another Mystery*
   - Have students enter their code and run their programs to check their homework
4. Review the three-dimensional coordinate system
   - Set up a coordinate system in the classroom
   - Have students build small models of the coordinate system in their groups
   - Review that the graph of a one-variable linear equation is a plane parallel to a coordinate plane
   - Have students visualize the polyhedron formed by a given set of vertices
5. *A Return to the Third Dimension*
   - Students review basics about the three-dimensional coordinate system
   - No whole-class discussion is needed

At Home

Homework 23: Where's Madie?

Special Materials Needed

- Three large (5" × 8" or larger) index cards (or pieces of tag board) per group
- Yarn and a pair of scissors for each group
- A transparency of the diagram showing the three-dimensional coordinate system (see Appendix C)
- (Optional) String for making a large-scale three-dimensional coordinate system, with the origin in the center of the classroom

Day 23

1. Forming New Groups

This is an excellent time to form new random groups. Follow the procedure described in the IMP *Teaching Handbook,* and record the members of each group and the suit for each student.

2. Reminder on POW Descriptions

Remind students that tomorrow they should turn in a general description of what they expect their animation programs will do.

3. Discussion of *Homework 22: Another Mystery*

Have students compare notes in their groups on the program to see if they can come to a consensus about what it does. Then have them enter the code they created and run the program to see what actually happens. If their predictions were wrong, let them take some time in their groups to try to figure out what went wrong with their analysis.

For appropriate window settings, the screen will look something like this:

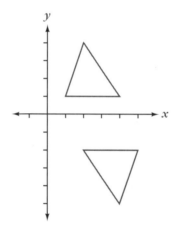

Students may be surprised that there are only two triangles, because the loop goes from $W = 1$ to $W = 4$. It turns out that when the triangle is rotated and translated a second time, it goes back to its original position.

Comment: Students may realize that the program involves a combination of a translation and a rotation. They may also see that the net effect of the two appears to be a rotation of the triangle around the point $(3, -0.5)$. If this comes up, you can tell them that any combination of a rotation and a translation is, in fact, another rotation, but not necessarily a rotation around the origin. Students may find this observation useful in working on *POW 5: An Animated POW.* The supplemental problem *The General Isometry* involves the process of combining translations, rotations, and reflections, and can be assigned after *Homework 24: Flipping Points.*

168 *As the Cube Turns* Interactive Mathematics Program

4. Reviewing the Three-Dimensional Coordinate System

The purpose of the next activity, *A Return to the Third Dimension*, is to refresh students' minds about the three-dimensional coordinate system and equations of planes in that system. As discussed in the Year 3 unit *Meadows or Malls?*, students generally find it helpful to use the classroom as a model of the three-dimensional coordinate system, and we recommend that you review this model before students begin the activity.

There are several ways to set up the classroom coordinate system (assuming that your classroom is a typical "box" shape). For consistency with the opening-day demonstration, we suggest that you use the blackboard as the xy-plane, with the z-axis coming out perpendicular to that plane.

The following alternative approach, suggested in *Meadows or Malls?*, allows you to consider both positive and negative coordinates within the classroom. First, choose a point in midair somewhere in the middle of the classroom. This point will represent the origin. Then tape three pieces of string across the room, through this point, to form straight lines in each of the three perpendicular directions (floor to ceiling, front to back, left to right). These pieces of string will represent the axes. Be sure to label the strings as the x-axis, y-axis, and z-axis, and to assign one end of each axis as the positive direction.

We suggest that you have the x-axis connect the side walls, the y-axis connect the floor and ceiling, and the z-axis connect the front and rear walls. That way, the x-axis and y-axis are parallel to their usual blackboard positions in the two-dimensional coordinate system. Viewed from the back of the room, the situation looks roughly like this diagram, with the coordinate planes shown for clarity. (This diagram is included in Appendix C.)

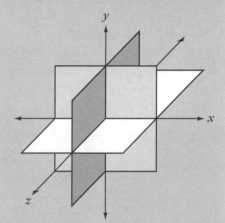

Note: In this setup, if students at the front of the room turn around to look at the x-axis, its positive direction will be to their left. This may be confusing, so be sure to identify the positive direction for each axis clearly.

Some teachers prefer to set up the axes with the origin in a corner on the floor of the classroom. In that approach, points in the classroom have positive coordinates.

Once you establish the setup of the axes, you can continue the review by having different students locate various ordered triples as points in the room and then find the coordinates of points from their location.

- **Building a small model**

    Question 1 of today's activity, *A Return to the Third Dimension,* suggests that students use a small-scale model of the coordinate system to understand what's going on. Many students find it helpful to have such a model, which they can build using index cards as coordinate planes. We suggest that each group build such a model now and save it for use over the rest of the unit.

    To do this, have each group start with two full-size index cards and two half cards, with slits cut as shown here by the dotted lines.

    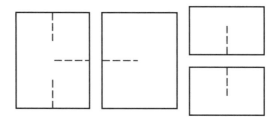

    They should fit the first two cards together along the slits shown here as horizontal, keeping the first card vertical and turning the second card to lie horizontal. They should then fit the two smaller cards so their slits match those at the top and bottom of the first card, turning the smaller cards so that they are perpendicular to both of the larger cards.

    Once students have constructed these models, you can have them practice locating points within this system.

- **Graphs of one-variable linear equations**

    "What is the graph of the equation $z = 5$ in this coordinate system?"

    Then ask what the graph is in this three-dimensional system for an equation such as $z = 5$. A good hint is to ask students to give you some ordered triples that satisfy the equation and to locate those points. They should see that they get a plane that can be described either as perpendicular to the z-axis or as parallel to the xy-plane.

    Then turn the process around, starting with a description of a plane parallel to one of the coordinate planes and asking the class for its equation. A good hint would be to ask students to name some points on the plane and then to think about what they have in common.

- **Identifying a polyhedron**

    As a final element in the review, ask students to imagine a polyhedron with four vertices that is placed in this coordinate system so that the vertices are located at (0, 0, 0), (4, 0, 0), (2, 0, 4), and (2, 4, 2). Give students a few minutes to work within their groups on visualizing this polyhedron.

"What does this polyhedron look like?"

Then ask for a description of what this looks like. Students should be able to articulate that this is a tetrahedron, in the first octant, and that its base is a triangle in the *xz*-plane. (If students do not know the term *tetrahedron*, you should introduce it here.)

Elicit such details as that one side of the base is along the *x*-axis, that the base is an isosceles triangle, and that the tetrahedron is 4 units "high," because vertex (2, 4, 2) at the "top" is 4 units directly above (2, 0, 2), which is a point of the base.

If more work along this line seems needed, you can ask similar questions requiring students to visualize objects in three dimensions.

5. A Return to the Third Dimension

After this introduction, turn students loose on the activity. You may not need to have a whole-class discussion of this—no further discussion ideas are provided here. Questions 2 and 3 should be considered optional and are intended for groups that finish Question 1 quickly.

Groups may find it useful to use yarn to outline the edges of their cube, as suggested in Question 1.

Homework 23: Where's Madie?

This assignment will give students more experience with the monorail idea, which will be extended to a three-dimensional situation through a series of activities beginning tomorrow.

A Return to the Third Dimension

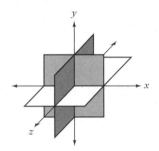

In the Year 3 unit *Meadows or Malls?*, you worked with the three-dimensional coordinate system to understand and solve linear programming problems. This activity will help you review ideas about that coordinate system.

1. Imagine a cube in the three-dimensional coordinate system, placed in a position that you choose. You may want to use yarn or a small cube, together with a cardboard version of the coordinate system, to create a model of the situation.

   a. Write down the coordinates of the eight vertices of your cube.

   b. A cube has six faces. Give the equations of the six planes that contain the faces of your cube.

   c. Draw a sketch of your cube in the three-dimensional coordinate system.

2. Write the equations of several pairs of planes that are parallel to each other.

3. Write the equations of a pair of planes that are neither parallel nor perpendicular to each other. Sketch your two planes.

Where's Madie?

One day as he was pruning away, Clyde thought of something he wanted to ask Madie. He thought he'd take a break and walk over to see her.

He didn't remember where she was working, but he knew where his snack was on the line between them. He also remembered Madie saying she would have to walk twice as far for her snack that day as he had to.

1. If Clyde was at (2, 6) and the snack was at (12, 1), where was Madie?

2. Next, suppose that you don't know the exact coordinates of either Clyde's tree or the snack, but assume that the monorail still runs along the line $x = 12$. Create general instructions for finding the coordinates of the tree where Madie is working, and verify that your instructions work for the situation in Question 1.

3. Let Clyde's coordinates be (a, b) and the snack's coordinates be $(12, c)$. Write a formula for the coordinates of Madie's tree.

And Fred Brings the Lunch

Mathematical Topics

- Continuing work in two dimensions finding points a certain fraction of the way along a line segment
- Finding points a certain fraction of the way along a line segment in three dimensions

Outline of the Day

In Class

1. Collect POW descriptions
2. Discuss *Homework 23: Where's Madie?*
   - Have students illustrate their reasoning with diagrams
3. *And Fred Brings the Lunch*
   - Students find points a fraction of the way between two points in three dimensions
   - The activity will be completed and discussed on Day 25

At Home

Homework 24: Flipping Points

1. Collect Descriptions for POW

Have students turn in their descriptions of what they expect to accomplish in *POW 5: An Animated POW.*

2. Discussion of *Homework 23: Where's Madie?*

Ask a volunteer to report on both Questions 1 and 2. For Question 1, the presenter might draw a right triangle using Clyde's and the snack's positions as two of the vertices (as shown on the next page), then extend the hypotenuse so the new portion is twice the length of the original

Interactive Mathematics Program   As the Cube Turns **175**

hypotenuse, and finally create a new right triangle with the new portion as the hypotenuse. The overall diagram might look something like the one shown here.

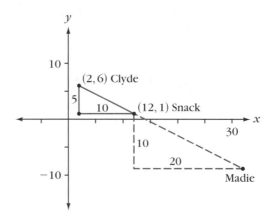

Elicit a clear explanation of how to use a diagram like this to see that Madie's tree is at $(32, -9)$.

Then have the same student present Question 2. As the presenter gives the general description, have the class apply it to the situation in Question 1 to verify that it matches what the student did in Question 1.

"Who did the problem in a different way?"

Next, ask for volunteers to describe different methods for answering Question 1 and then to give the instructions they developed in Question 2.

Finally, ask for volunteers to present a general formula. Students should get something equivalent to $(12 + 2(12 - a), c + 2(c - b))$. Have students verify that if $a = 2, b = 6$, and $c = 1$ (as in Question 1), they get the same result they got for Question 1. You may also want to have students test out the instructions developed for Question 2 to see if those instructions yield the same formula (or an equivalent one).

3. And Fred Brings the Lunch

Tell students that the next activity, *And Fred Brings the Lunch,* presents a problem similar to the orchard snack problems except that it involves three dimensions. You may want to suggest that students build a physical model of the situation using cardboard. Or they might visualize the situation in terms of their classroom coordinate system.

You can ask each group to prepare a presentation on one of the questions. The activity is scheduled to be discussed on Day 25.

Homework 24: Flipping Points

> This homework will give students more experience working with coordinate geometry. This material is not needed for solving the unit problem.

And Fred Brings the Lunch

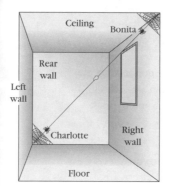

Once there were two spiders, Bonita and Charlotte. They liked to spin webs at opposite corners of a room. Bonita liked heights, and she always went to one of the corners of the ceiling.

Charlotte did not care for heights, and she always went for a corner of the floor. To give the proper artistic balance to the room, she went to the corner diagonally opposite from Bonita.

The spiders' webs were connected by a single thread. As they worked on their webs, they got hungry. So they had their associate, Fred the fly, bring in lunch each day. The spiders wanted their lunch partway between them along the connecting thread.

Fred did not glide up or down threads as the spiders did. He simply flew places. So Bonita and Charlotte mapped out a coordinate system to tell him where to fly. They designated the rear, left corner of the floor, where Charlotte usually hung out, as (0, 0, 0).

Bonita was in the front, right corner of the ceiling. The spiders set up the system so that the coordinates for her position, (a, b, c), worked this way.

- The first coordinate, a, represented how many feet to the right Bonita was from Charlotte.

- The second coordinate, b, represented how many feet up Bonita was from Charlotte.

- The third coordinate, c, represented how many feet forward Bonita was from Charlotte.

Continued on next page

1. One day, the spiders were in a room that was 11 feet wide, 13 feet tall, and 24 feet long. If Charlotte was at (0, 0, 0) and Bonita was at the opposite corner at (11, 13, 24), and they wanted lunch halfway between them, where should Fred bring lunch?

2. After she'd gotten used to this room, Charlotte got adventurous and built a web slightly out from her corner, at (2, 1, 4). Bonita also left her corner, building her web at (9, 12, 20). The connecting thread still went in a straight line between Bonita's web and Charlotte's, and they still wanted lunch halfway between them. Where should Fred make his delivery?

3. Charlotte suddenly decided that with her distaste for heights, she should not have to go halfway up. She and Bonita compromised, agreeing that Charlotte would go one-third of the way up the thread from her position while Bonita would come two-thirds of the way down. Based on Bonita's and Charlotte's positions in Question 2, where should Fred go?

4. Using the positions from Question 2 again, where should Fred go if the lunch is to be placed two-fifths of the way from Charlotte to Bonita?

5. Charlotte and Bonita are tired of all this mental activity, figuring out what to tell Fred. Could you provide a formula?

    Assume that Charlotte is at (x_1, y_1, z_1), that Bonita is at (x_2, y_2, z_2), and that r is a fraction, so that they want lunch "r of the way" along the thread from Charlotte to Bonita. Create a formula in terms of these variables that states where Fred should bring lunch.

Flipping Points

You have learned about translations and rotations, and developed formulas to find the new coordinates after one of these geometric transformations has been applied to a point.

Translations and rotations are two important examples of a special kind of transformation called an **isometry.** An isometry is a way of moving all the points in the plane (or in 3-space) so that the size and shape of objects are unchanged. The word *isometry* means "same measure," which means that the distance between two points doesn't change when the points are moved.

There is a third basic category of isometry called a **reflection.** (Reflections are also known as *flips*.) A reflection in the plane is defined by giving a **line of reflection.** The reflection then moves each point P to the point Q so that the line of reflection becomes the perpendicular bisector of the segment connecting P and Q. In other words, Q is chosen so that $\overline{PQ}$ is perpendicular to the line of reflection and the line of reflection intersects $\overline{PQ}$ at its midpoint. Point Q is called the reflection of P across the line of reflection.

For example, in the diagram at the left, L is the line of reflection, point Q is the reflection of point P across L, and L is the perpendicular bisector of $\overline{PQ}$. Thus, the reflection of a point is the "mirror image" of the original point through the line of reflection.

Continued on next page

The diagram below shows a triangle in the first quadrant and its reflection using the *y*-axis as the line of reflection.

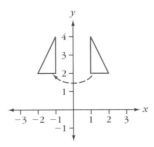

For instance, the reflection of the point (1, 2) is the point (−1, 2), and the line of reflection (the *y*-axis) is a line of symmetry between the original triangle and its reflection.

1. Using the diagram, give the coordinates for each of the other two vertices of the original triangle (in the first quadrant) and give the vertices of each of their reflections (in the second quadrant).

2. Generalize the result of Question 1. That is, if (a, b) is an arbitrary point, what are the coordinates of its reflection through the *y*-axis?

3. Find a way to represent this transformation using a matrix.

DAY 25

Students apply similar triangles to three dimensions.

Fred Brings More Lunch

Mathematical Topics

• Expressing reflections using coordinates and using matrices
• Continuing to find points a certain fraction of the way along a line segment in three dimensions

Outline of the Day

In Class
1. Discuss *Homework 24: Flipping Points*
   • Students share their work
2. Complete and discuss *And Fred Brings the Lunch*
   • Have students justify their work using similar triangles

At Home
Homework 25: Where's Bonita?

1. Discussion of *Homework 24: Flipping Points*

Let club card students report on Questions 1 and 2. These should be fairly straightforward, with students seeing that the reflection of the point (a, b) is the point $(-a, b)$. You may want to bring out explicitly that this is true even if a or b is already negative.

Ask for a volunteer on Question 3. If no one was able to answer this question, tell students to try matrix multiplication. If they need a more explicit hint, tell them to look for a matrix B such that $[a \ b] B = [-a \ b]$. If necessary, set up B as $\begin{bmatrix} r & s \\ t & u \end{bmatrix}$. Students should see that they want $B = \begin{bmatrix} -1 & 0 \\ 0 & 1 \end{bmatrix}$.

Comment: The supplemental problem *The General Isometry* asks students to show that every isometry in two dimensions can be obtained as a combination of a translation, a rotation, and, if needed, a reflection.

Day 25

2. Discussion of *And Fred Brings the Lunch*

> If necessary, give students additional time to complete the activity and prepare presentations.

The individual problems in *And Fred Brings the Lunch* are all three-dimensional analogs of the orchard snack problems. You can begin by having different spade card students present Questions 1 through 4.

On Questions 1 and 2, students can adapt the midpoint formula to three dimensions to get the lunch locations as $\left(\frac{0+11}{2}, \frac{0+13}{2}, \frac{0+24}{2}\right)$ and $\left(\frac{2+9}{2}, \frac{1+12}{2}, \frac{4+20}{2}\right)$ respectively.

For Questions 3 and 4, students will need to adapt to three dimensions the generalization of the midpoint formula that they developed on Day 21 in the activity *Fractional Snacks* (or reconstruct the reasoning used there).

For example, in Question 3, Bonita is 7 units to Charlotte's right, 11 units above her, and 16 units forward from her. Because the lunch is one-third of the way from Charlotte to Bonita, it should be placed $\frac{7}{3}$ units to Charlotte's right, $\frac{11}{3}$ units above her, and $\frac{16}{3}$ units forward from her. This puts the lunch at $\left(4\frac{1}{3}, 4\frac{2}{3}, 9\frac{1}{3}\right)$.

Similarly, in Question 4, the lunch should be located at the point $\left(2 + \frac{2}{5} \cdot 7, 1 + \frac{2}{5} \cdot 11, 4 + \frac{2}{5} \cdot 16\right)$, which simplifies to $\left(4\frac{4}{5}, 5\frac{2}{5}, 10\frac{2}{5}\right)$.

- **Question 5**

    Ask for a volunteer for Question 5, in which r represents some fraction of the way we want to go from (x_1, y_1, z_1) to (x_2, y_2, z_2). Students should see that the lunch should be located at

    $$\left(x_1 + r(x_2 - x_1), y_1 + r(y_2 - y_1), z_1 + r(z_2 - z_1)\right)$$

    and should recognize that this is a three-dimensional version of the formula they developed for *Fractional Snacks*.

- **Why does it work?**

    "How do you know that the reasoning you used in the two-dimensional case works here?"

    You should ask students how they know that the reasoning they used in the two-dimensional orchard snack problems is applicable here. If a hint is needed, ask them to think back to the two-dimensional case and redevelop the reasoning used there. They should see that the argument was based on similar triangles.

    Similar triangles also work for the three-dimensional case, but the reasoning involves a two-stage process and is more difficult to visualize. You can be satisfied if students simply recognize that similar triangles are the necessary concept here, without getting into all of the details.

Day 25

Students may also argue that "of course" you can simply add a third coordinate to the formula. If this comes up, point out the difference between something seeming reasonable and something being proved.

Homework 25: Where's Bonita?

This assignment is a variation on *And Fred Brings the Lunch* and is a three-dimensional version of *Homework 23: Where's Madie?* This homework can be assigned even if students have not yet gotten the general formula from *And Fred Brings the Lunch*.

Where's Bonita?

Fred generally got his instructions from Charlotte, and so he knew what her coordinates were. Of course, he also knew where he was putting the lunch. But after a while, he started wondering where Bonita was.

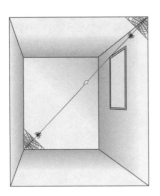

1. For example, one day, when Charlotte was at (1, 3, 2), she told Fred that the lunch should be at (5, 8, 9) and that this was halfway to Bonita. Where was Bonita?

2. On another occasion, when Charlotte would only go one-third of the way toward Bonita to get lunch, Charlotte told Fred to put the lunch at (4, 4, 7). Charlotte herself was at (2, 1, 1). Where was Bonita?

3. Fred's brain was really taxed one day when Charlotte told him to put the lunch at (3, 6, 7), and explained that this would be two-fifths of the way to Bonita. Charlotte was at (1, 2, 2) when she said this. Where was Bonita?

DAY 26

Lunch in the Window

Students find the point where a given plane intersects a given line.

Mathematical Topics

- Finding the point where the segment connecting two given points meets a given plane
- Projecting from three dimensions to two dimensions

Outline of the Day

In Class

1. Discuss *Homework 25: Where's Bonita?*
2. *Lunch in the Window*
   - Students find the point where a given vertical plane meets a line segment in 3-space
   - Have students begin work and bring them together for discussion of Question 1 when groups have finished it
   - After discussing Question 1, have students return to the activity
3. Discuss *Lunch in the Window*
   - After discussing Question 2, bring out the similarity of this problem to the monorail situation
   - Post a general formula for the situation

At Home

Homework 26: Further Flips

1. Discussion of *Homework 25: Where's Bonita?*

Give groups only a few minutes to compare ideas, and then let heart card students report on each problem. Be sure to have them explain how they got their answers.

For instance, in Question 1, Charlotte goes 4 units to the right, 5 units up, and 7 units forward to get from her position, (1, 3, 2), to Fred's position, (5, 8, 9). Because this point is halfway to Bonita, Bonita must be 4 units to the right, 5 units up, and 7 units forward from (5, 8, 9), which puts her at (9, 13, 16).

Day 26

Students might explain this by saying that to get from Charlotte to Fred, you "add" (4, 5, 7), and so you add the same amount to get from Fred to Bonita.

For Question 2, students need to see that if Fred is one-third of the way from Charlotte to Bonita, then the distance from Fred to Bonita is twice as far (in each of the three directions) as that from Charlotte to Fred.

There is no need to develop a general formula for this process.

> *Note:* In Question 3, if the fact that the fraction is not a unit fraction caused difficulties, you can give a hint such as, "What is the *x*-coordinate of the point that is one-fifth of the way from Charlotte to Bonita?"

2. *Lunch in the Window*

We suggest that you have students read the introduction to *Lunch in the Window* and work in groups on Question 1. Then bring the class together to discuss that part of the assignment.

"What previous activity is this like? In what way?"

You can begin the discussion by bringing out the similarity between this assignment and *Monorail Delivery* (Day 22), perhaps first asking what previous activity *Lunch in the Window* resembles. The class should see that the plane of the window is analogous to the line along which the monorail ran.

Then turn to Question 1 itself. Students should recognize that all points in the plane containing the window are 14 feet to the right of the origin, so this is the plane $x = 14$.

"What does the equation of the plane tell you about the coordinates of the lunch position?"

Next, ask students what that equation tells them about the coordinates of the point where the lunch should be located. They should be able to state that the lunch position must be a point whose *x*-coordinate is 14.

If a hint is needed, ask:
"What part of the distance does Fred Jr. have to go in the x-direction?"

Once Question 1 has been discussed, have groups resume work on the activity. If they are stuck on Question 2, you can ask what fraction of the distance (from Charlotte to Bonita) Fred Jr. must travel in the *x*-direction.

When all groups have finished Question 3, you can begin the discussion. (If no group has finished Question 5, you can let groups work on that after discussing the earlier problems.)

3. Discussion of *Lunch in the Window*

For Question 2, many students probably will have noticed that 14 (Fred Jr.'s *x*-coordinate) is halfway between 4 and 24 (Charlotte's and Bonita's *x*-coordinates), so the window must be halfway between them in terms of left/right position. The reasoning used in *And Fred Brings the Lunch* shows that the lunch spot must also be halfway in the other two directions. In other words, the desired lunch spot is exactly at the midpoint between Charlotte and Bonita, which is (14, 11, 5).

You may want to bring out again the analogy between this problem and *Monorail Delivery*. Students should recognize that in both problems, they are given one of the coordinates of the desired point and that the key step is using that coordinate to find the value of r.

In Question 3, they can use similar reasoning, although the arithmetic is less obvious. Here, the two spiders' x-coordinates are 10 and 26. Students with good number sense will have seen that the desired point in the window must be one-fourth of the way from Charlotte to Bonita, because 14 is one-fourth of the way from 10 (Charlotte's x-coordinate) to 26 (Bonita's x-coordinate). And because the desired point is one-fourth of the way from Charlotte to Bonita in the x-direction, it must be one-fourth of the way in all directions.

"How could you find the value of r computationally?"

Ask students how they might come up with the value $r = \frac{1}{4}$ computationally. They should be able to identify this as the ratio $\frac{14 - 10}{26 - 10}$.

It may help to introduce the term **x-distance** (or *x-component of the distance*) to describe the difference in x-coordinates between two points. Using this term, students might express the computation like this:

"Because the lunch spot is one-fourth of the way from Charlotte to Bonita in terms of x-distance, it should also be one-fourth of the way from Charlotte to Bonita in terms of y- and z-distance."

Once they have found r, students can either use yesterday's formula

$$(x_1 + r(x_2 - x_1), y_1 + r(y_2 - y_1), z_1 + r(z_2 - z_1))$$

substituting $\frac{1}{4}$ for r, or work more intuitively.

If time is short, you do not need to discuss Question 4. The principle here is the same, but the numbers are a bit messier.

- **Question 5: A general rule for Fred Jr.**

    Next, turn to the generalization called for in Question 5. (If no group got to this question yet, you can let groups work on it now.) Students should come up with something equivalent to $\frac{14 - x_1}{x_2 - x_1}$ as the value of r, that is, as the ratio between the distance from Charlotte to the lunch spot and the distance from Charlotte to Bonita.

    The expression $r(y_2 - y_1)$ says how far above Charlotte the lunch spot is (that is, the y-distance from Charlotte to lunch), so the y-coordinate of the lunch spot is $y_1 + \left(\frac{14 - x_1}{x_2 - x_1}\right)(y_2 - y_1)$. Similarly, $r(z_2 - z_1)$ says how far forward from Charlotte the lunch spot is, so the z-coordinate of the lunch spot is $z_1 + \left(\frac{14 - x_1}{x_2 - x_1}\right)(z_2 - z_1)$.

Day 26

You can help students confirm this reasoning by asking what the algebra pattern in these formulas would give as the x-coordinate of the lunch spot. Students should see that the analogous expression is $x_1 + \left(\frac{14 - x_1}{x_2 - x_1}\right)(x_2 - x_1)$, which simplifies to $x_1 + (14 - x_1)$, which is equal to 14. They should see that this is what it ought to be, because the lunch spot is in the plane containing the window, whose equation is $x = 14$.

Finally, have students generalize to the case where the window is in the plane $x = k$. You can then post a summary of the formulas, which might look like this:

The line connecting the points (x_1, y_1, z_1) and (x_2, y_2, z_2) meets the plane $x = k$ at the point

$$(k, y_1 + r(y_2 - y_1), z_1 + r(z_2 - z_1))$$

where r is the ratio $\frac{k - x_1}{x_2 - x_1}$.

Homework 26: Further Flips

This assignment continues the ideas of *Homework 24: Flipping Points*.

Lunch in the Window

The two spiders are still spinning away, but Bonita has developed claustrophobia and prefers to work outside. So the spiders found a room where one of the windows is kept open. Charlotte works inside (which she prefers), while Bonita does her spinning outside.

The thread connecting them goes through the open window, and Charlotte and Bonita will have their lunch right where the thread passes through the opening of the window. Therefore, they need to figure out the coordinates of the point where the thread passes through the open window.

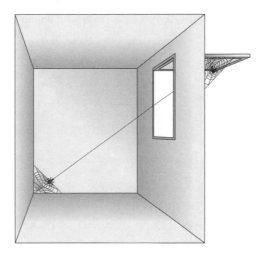

Continued on next page

Charlotte and Bonita are using the same coordinate system as before, with the origin at the rear, left corner of the floor. The first coordinate gives distance to the right, the second gives distance up, and the third gives distance toward the front. The window is in the wall on the right side of the room, and the room is 14 feet wide from left to right.

By the way, the spiders' associate Fred mysteriously disappeared after the last lunch. His son, Fred Jr., is now assisting them.

1. Before you get started on the lunch problems, find the equation of the plane that contains the window.

2. Suppose that Charlotte is at (4, 2, 2) and Bonita is at (24, 20, 8). Where should Fred Jr. bring their lunch? Explain your reasoning carefully.

3. Now suppose Charlotte is at (10, 0, 4) and Bonita is at (26, 16, 8). Where should Fred Jr. drop off lunch? Again, explain your reasoning.

4. This time Charlotte is at (7, 0, 4), while Bonita is still at (26, 16, 8). Give Fred Jr. directions about where to put lunch.

5. Generalize your results, based on the coordinates (x_1, y_1, z_1) for Charlotte and (x_2, y_2, z_2) for Bonita.

Further Flips

In *Homework 24: Flipping Points,* you looked at the isometry of reflecting figures through the *y*-axis, as shown in the first diagram here. But any line can be used as the line of reflection. In this assignment, you will consider some other cases.

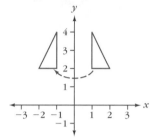

1. Start with the same original triangle, but this time use the line $y = x$ as the line of reflection, as shown in the next diagram. Find the coordinates of the vertices of both the original triangle and the reflected triangle.

2. Generalize Question 1 by finding the image of an arbitrary point (a, b) under the reflection through the line $y = x$.

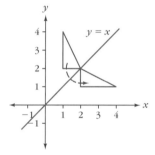

3. Express this reflection in terms of matrices. That is, find a matrix process that will turn the vector $[a \; b]$ into the corresponding image vector under the reflection.

4. Now repeat Questions 1 through 3 using the same original triangle and using the line $x = 6$ as the line of reflection. (*Suggestion:* Draw the diagram.)

DAY 27

Cube on a Screen

Students draw the projection of a cube on a screen to get a feel for the problem.

Mathematical Topics

• Continuing work with reflections in the plane and expressing them using matrices

• Using a physical situation to draw the projection of a cube onto a plane

Outline of the Day

In Class

1. Discuss *Homework 26: Further Flips*
   • Bring out that not every reflection can be achieved through matrix multiplication

2. *Cube on a Screen*
   • Students use their line of sight to draw the projection of a cube on a plane
   • When students are paired and have the materials, demonstrate the task
   • (Optional) Bring the class together for a discussion of Part I before pairs work on Part II

3. Discuss *Cube on a Screen*
   • Emphasize that the drawings are actually different, and not simply translations or rotations of one another
   • Discuss how the activity is related to the unit problem

4. Have students begin thinking about how to represent projection in terms of coordinates

At Home

Homework 27: Spiders and Cubes

Special Materials Needed

• Sheets of Plexiglas or other firm, clear material (one sheet for each pair of students)
• Pens of three different colors for writing on the firm, clear material (a set of three pens for each pair of students)
• Cubes (one per pair, at least 2 inches on an edge, preferably made by connecting smaller cubes with different colors at the corners)

1. Discussion of *Homework 26: Further Flips*

The initial situation in Questions 1 through 3, involving reflection through the line $y = x$, should be fairly straightforward. Students should see (perhaps using the vertices of the triangle as a model) that the image of (a, b) is (b, a).

Interactive Mathematics Program   As the Cube Turns   193

Based on their experience with *Homework 24: Flipping Points*, students will probably know to look for a matrix M such that $[a \ b] M = [b \ a]$. A little experimentation should show them that they want $M = \begin{bmatrix} 0 & 1 \\ 1 & 0 \end{bmatrix}$.

For Question 4 (reflection of the original triangle through the line $x = 6$), students may have some difficulty. With the use of a diagram, students should be able to get the images of the individual vertices:

$$(1, 2) \longrightarrow (11, 2)$$
$$(2, 2) \longrightarrow (10, 2)$$
$$(1, 4) \longrightarrow (11, 4)$$

If needed, suggest that students try other points and make an In-Out table, focusing in particular on the *x*-coordinates. They should see that the image of (a, b) is $(12 - a, b)$. But they may have to struggle to reach the conclusion that this transformation can't be achieved by matrix multiplication. You can either leave this as an open question or suggest that they look at *Homework 22: Another Mystery* for ideas.

> The transformation in *Homework 22: Another Mystery* was not a reflection, but it did illustrate the idea of combining basic transformations. (Specifically, it combined a rotation around the origin with a translation.) The reflection in Question 4 of last night's assignment can be achieved as a combination of a reflection through a line through the origin and a translation. It can be represented using a combination of matrix multiplication and matrix addition.
>
> As noted previously, the supplemental problem *The General Isometry* asks students to show that every isometry in two dimensions can be obtained as a combination of a translation, a rotation, and, if needed, a reflection.

2. Cube on a Screen

> In movies with animation such as *Star Wars*, we often see objects zooming by as we look out the window of the moving spaceship. To make this look realistic, the animator must change the object's shape to reflect the fact that the viewpoint is changing or that the distance between the viewer and the object is changing. The purpose of having student pairs do several drawings in this activity is for them to see that the drawings will look different.

In tomorrow's activity, *Find Those Corners!*, students will apply the concepts from *Lunch in the Window* to the unit problem, specifically to item 4: "Create a two-dimensional drawing of a three-dimensional object."

Day 27

Today's activity, *Cube on a Screen,* is a concrete introduction to the situation of tomorrow's more computational activity. It gives students a chance to carry out in detail the demonstration you did on Day 20 (see the section "Introduction to Projections").

- ### Reviewing the motion of the cube

    Give each pair of students a large colored cube and a sheet of Plexiglas (or similar material) and three pens of different colors. If possible, have students create this large cube from small colored cubes so that there are different colors at the corners. (The colors will help them keep track of their work.)

"What's the unit problem?"

Ask students to summarize the central unit problem, which is to show a rotating cube on a calculator or computer screen. You may want to review the roles and positions of the three axes (see the diagram on Day 1). Remind the class, as needed, that the x- and y-axes are the horizontal and vertical axes in the plane of the wall or chalkboard, and that the z-axis is perpendicular to this xy-plane. Bring out that the cube is going around the z-axis and that it makes one complete spin as it goes around.

Have students work in pairs to act out the motion using their cubes. One can hold the Plexiglas, which represents the screen, while the other turns the cube from behind the screen.

- ### Beginning the activity

    Have students read through the instructions as a class, or explain the sequence of tasks to them. Because the directions are complex, you can demonstrate what to do, perhaps reviewing the demonstration you did on Day 20. In particular, you may want to demonstrate what is meant in Question 6 by the statement "partner B moves the cube through a partial rotation (such as 45° or 90°) around the 'z-axis.'"

    Once students begin, you can circulate and make sure that they have the right picture of what is happening. We suggest that you discuss Part I as soon as all pairs have completed it, and then let them go back to work on Part II.

    > If partner B moves only slightly between Questions 1 and 2, or if partner A moves the cube only slightly between Questions 2 and 3, the differences may not be readily apparent in the tracing. If this point is not clear, you may want to have students make a more dramatic change in position so they do see different faces.

    Note: Tonight's homework does not depend on the discussion of this activity, so if necessary, you can complete the discussion tomorrow.

Day 27

3. Discussion of *Cube on a Screen*

- *Part I: Changing the viewpoint, moving the cube*

    After all pairs have completed Part I, ask for volunteers to comment on how these sketches compared with one another. Students should see that the projection changes if either the observer or the cube is at a different position. For example, when the observer moves, she or he may see different faces of the cube. In particular, students should see that the drawing is larger when the cube is closer to the screen. This activity should sharpen students' intuition about how the picture changes as distances change.

"How is the fact that the picture changes when the viewpoint changes related to the unit problem?"

Ask students how the fact that the picture changes when the viewpoint changes is connected to the unit problem. They should see that because of this fact, they will need to decide on a specific location for the observer, in relation to the screen and the object, when they write a program for the unit problem. That location is called the **viewpoint** or the **center of the projection.**

- *Part II: Rotating the cube*

    When most pairs have completed Part II, ask for volunteers to comment on their results.

    Students should see that the drawing of the rotated cube is *not* simply a rotation of the original drawing. For example, they should see that they cannot simply turn partner A's drawing from Question 5 and make it match up with the drawing from Question 6.

    This point is somewhat subtle, but it is very important for the development of the final program for the unit. Therefore, you should take care to ensure that students appreciate the distinction.

    For example, in addition to having students see that these two drawings are not simply rotations of each other, you can take some particular drawing of a cube (on either Plexiglas or paper), hold it up in front of the class, and then turn the drawing. Ask students if this is what a cube rotating in three dimensions should look like. You may want to then have students go back to their actual drawings to bring out that simply rotating a drawing does not give the right result.

"What needs to rotate in the unit problem—the cube or the picture on the screen?"

As with Part I, ask the class how the fact that the picture does more than simply rotate as the cube rotates affects the unit problem. Help students to see that this means that when they write their programs, they must create a new projection for each small rotation of the cube. In other words, they will set up their cube in its initial position, draw the projection, then rotate the cube and find the coordinates of the vertices of the rotated cube, draw the projection of that new cube, and so on. *Important!* Students need to realize that it is not correct simply to make one projection and then turn that two-dimensional picture.

4. Projection Using Coordinates: An Overview

Ask students now to think about the projection process that they have done visually with the Plexiglas, and imagine doing this abstractly in terms of a three-dimensional coordinate system. That is, they should contemplate this scenario:

- Instead of having a physical cube, they have been given the *coordinates* for each of the corners of the cube.

- Instead of placing their head in some position and looking at the cube, they have been given *coordinates* for the viewer's eye.

- Instead of having a sheet of Plexiglas to represent the screen, they have been given *an equation* for the screen.

"Given only this information, how would you determine where on the screen to draw the corner of the cube?"

Ask them to consider the task of figuring out, from this information, where on the screen they would draw each corner of the cube. They will actually be doing the arithmetic and algebra of this process in tomorrow's activity, *Find Those Corners!*

"How does this problem relate to 'Lunch in the Window'?"

For now, simply have students contemplate it briefly. You may want to start them thinking about their homework reflection by asking how this process is related to the *Lunch in the Window* problem.

> *Comment*: In the sixteenth century, some artists figured out where to put a point on a canvas by putting the canvas itself on hinges between themselves and the object they were painting, and then swinging the canvas out of the way. Then they ran a string from the object they were drawing to their eye and figured out where that string would have gone through the canvas. Tomorrow, students will learn mathematical techniques by which to find the spot.

Homework 27: Spiders and Cubes

> This assignment asks students to reflect on how the last few days relate to the main unit problem.

Interactive Mathematics Program

As the Cube Turns

Day 27

Cube on a Screen

As part of the unit problem, you need to figure out how to draw a cube, which is three-dimensional, on the calculator screen, which is two-dimensional. Such a drawing is called a **projection.**

This activity should help you get some insight into how this can be done. You should work with a partner on this activity. You and your partner will need a cube, a sheet of clear plastic, and three pens of different colors. You can think of the plastic as representing the calculator screen. It should be set up vertically like a plane parallel to the xy-plane.

In Part I, partner A will hold the cube and screen while partner B makes three different sketches of the cube, as described in Questions 1, 2, and 3. In Question 4, the two partners will compare the three sketches.

The partners will switch roles in Part II, and partner A will make two sketches, as described in Questions 5 and 6. In Question 7, the two partners will again compare the sketches.

Part I: Changing the Viewpoint, Moving the Cube

1. To begin with, partner A should take the cube and the screen, and place them in some fixed position, with the screen held vertically between the cube and partner B. Partner B need not be directly in front of the cube.

Continued on next page

Without moving her head, partner B should look at one corner of the cube and place a dot on the screen where her line of vision crosses the screen. (Partner B may want to imagine a laser beam from her eye to the corner of the cube. Place the dot where the beam would burn a hole in the screen.)

Partner B should continue like this, without moving, imagining her line of sight tracing all the edges of the cube, and marking on the screen where her line of vision would cross the screen as the cube is traced. The result should be a two-dimensional drawing—a projection—of the three-dimensional cube.

2. When the drawing from Question 1 is complete, partner B should move so that she is in a different position compared to the cube. *The cube and screen should be left in the same position as in Question 1.*

Partner B should now do a tracing of the cube from this new position, using the same "laser" method as in Question 1, but with a pen of a different color.

Continued on next page

3. Next, partner A should move the cube closer to the screen. *Partner B should stay in the same position as in Question 2.* Partner B should make a third sketch of the cube, again using the "laser" method, but with the third pen.

4. The two partners should compare the drawings from Questions 1 through 3. How are they different? Has the drawing merely moved, or has it actually changed?

Part II: Rotating the Cube

Now the partners should switch roles, with partner A doing the drawing and partner B holding the cube. Hold the screen so that a clean portion of it is between partner A and the cube.

5. Partner A should make an initial sketch of the cube, as described in Question 1.

6. Partner A and the screen should stay in the same position while partner B moves the cube through a partial rotation (such as 45° or 90°) around the "z-axis." That is, imagine a line perpendicular to the screen, and have the cube do a partial rotation around this axis, turning as it goes around the axis as in the central unit problem.

   Partner A should do a sketch of the cube with the cube in its new position, using a pen of a different color from that used in Question 5.

7. The two partners should then compare the drawings from Questions 5 and 6. How are they the same, and how are they different? Has the drawing merely turned, or has it actually changed?

Spiders and Cubes

You've recently completed a series of problems involving the eating habits of the spiders Bonita and Charlotte.

- *And Fred Brings the Lunch*
- *Homework 25: Where's Bonita?*
- *Lunch in the Window*

But the unit is about programming a graphing calculator to draw a turning cube. What do lunch arrangements for spiders have to do with the unit problem? Your task in this assignment is to figure out and explain this connection.

DAY 28

Find Those Corners!

Mathematical Topics

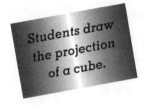

- Connecting recent activities with projection from three dimensions to two dimensions
- Expressing projections in terms of coordinates

Outline of the Day

In Class

1. Remind students that POW outlines are due tomorrow
2. Discuss *Homework 27: Spiders and Cubes*
   - Have students review the connections between *Lunch in the Window*, *Cube on a Screen*, and the unit problem
3. *Find Those Corners!*
   - Groups are given different viewpoints by the teacher, and each group projects a cube onto a screen based on its viewpoint
   - Students will continue the activity on Day 29, and discussion will be on Day 30

At Home

Homework 28: An Animated Outline

Special Materials Needed

- Large sheets of graph paper (at least one per group)
- A cube for each group
- The three-dimensional coordinate systems students made from index cards on Day 23

1. Reminders on *POW 5: An Animated POW*

Remind students that their outlines for their POWs are due tomorrow. They will complete these outlines in tonight's homework assignment.

Day 28

2. Discussion of *Homework 27: Spiders and Cubes*

Let students share ideas as a whole class about the connection between the spider problems and the unit task of programming a turning cube on a screen.

Review yesterday's activity, *Cube on a Screen,* and the discussion of how it relates to the situation in *Lunch in the Window*. To summarize:

- The viewpoint and the corner of the cube are like the two spiders.
- The line from the viewpoint to the corner is like the thread connecting the spiders.
- The screen is like the window.
- Figuring out where to place the dot on the screen is like finding the place where the thread goes through the window.

Bring out as well that just as *Lunch in the Window* can serve as a model for understanding *Cube on a Screen,* so also *Cube on a Screen* is itself a model for the process of drawing a cube on the calculator screen. In other words, students can think about the problem on three levels:

- The spider model
- The drawing on the clear screen
- The drawing on the calculator screen

3. *Find Those Corners!*

Tell students that they are finally ready to do the numerical analysis involved in carrying out item 4: "Create a two-dimensional drawing of a three-dimensional object." You can tell them that once they know how to make a two-dimensional projection of a stationary cube, they will focus on rotating the cube and then finishing the unit problem.

As groups begin work on *Find Those Corners!,* you should assign a specific viewpoint to each group (see the list in the subsection "Suggested viewpoints and their projections"). Each group should make a poster on grid chart paper of its projected cube. If any groups finish early, you can assign another viewpoint.

To save on chart paper, you can ask students to first do their work on their own paper. When they are satisfied with their work, they can transfer it to grid chart paper.

- *Suggestions for students*

    Because students will be comparing the results of different groups, it will be helpful to have all groups use the same scale. Based on the suggested

Day 28

viewpoints listed in the next subsection, 1-inch grid paper works well, with 2 inches per unit on both axes and both axes going from −2 to 4.

Suggest to students that they do their front face first. Once they complete that face, they can work with the back face. They can label each two-dimensional projection with the three-dimensional coordinates of the corner of the cube that it represents. Once they have drawn a projection point, they should connect it to all the other corner points on a common edge.

You might also suggest that at least two students in each group should independently find the coordinates of each projection and that they compare results before the group plots the point.

- *Suggested viewpoints and their projections*

    Here are some viewpoints that work well with the scales just described:

    - (4, 5, 10)
    - (1, 1, 10)
    - (1, 5, 10)
    - (4, −3, 10)
    - (−3, 5, 10)

    - (−3, −4, 10)
    - (1, −3, 10)
    - (3, 1, 10)
    - (4, 5, 20)
    - (1, 1, 6)

(Using 10 as the z-coordinate of the viewpoint makes the arithmetic come out nicely.)

You should post the list of viewpoints that you assign. When the activity is completed, students will post their diagrams showing the projected cubes and try to match each poster with its associated viewpoint.

As groups work on this, you can circulate and give help as needed.

For teachers: An outline of the process

For your convenience and reference, we describe here a sequence of steps for finding the projection of the vertex at (2, 2, 2) from the viewpoint (4, 5, 10), with the screen at $z = 5$, using the same reasoning as in *Lunch in the Window*. (The same screen location is used in the activity for all viewpoints.)

Note: Here, we are thinking of the projection as being part of the way from the vertex to the viewpoint. We could just as well think of it as being part of the way from the viewpoint to the vertex. In that case, the value of r would be $\frac{5}{8}$ instead of $\frac{3}{8}$, and the roles of the two points would be interchanged.

Students' first step will probably be to see that the z-distance from the vertex at (2, 2, 2) to the screen is 3 units and that the z-distance from (2, 2, 2) to the viewpoint (4, 5, 10) is 8 units. They should conclude that the projection of (2, 2, 2) is $\frac{3}{8}$ of the way from (2, 2, 2) to (4, 5, 10). In other words, they will find that for this case, the number represented as r in the spider lunch and orchard snack problems comes out to $\frac{3}{8}$. Therefore, the x- and y-distances from (2, 2, 2) to its projection must be $\frac{3}{8}$ of the corresponding distances from (2, 2, 2) to the viewpoint.

The x-distance from (2, 2, 2) to (4, 5, 10) is 2, and $\frac{3}{8}$ of this is 0.75. So the x-coordinate of the projection is 2.75 (adding the value 0.75 to the x-coordinate of the vertex). Similarly, the y-distance from (2, 2, 2) to (4, 5, 10) is 3, and $\frac{3}{8}$ of this is 1.125. So the y-coordinate of the projection is 3.125. Therefore, the projection of (2, 2, 2) is the point (2.75, 3.125, 5).

Help students to articulate that the ratio $\frac{3}{8}$ in this example corresponds to the r-value in the formula developed in the activity *And Fred Brings the Lunch*. More generally, this could be expressed as the ratio

$$\frac{z_{screen} - z_{vertex}}{z_{viewpoint} - z_{vertex}}$$

In drawing the cube on graph paper, students should realize that they need only look at the x- and y-coordinates of the projection. That is, the graph paper is assumed to represent the plane $z = 5$, so points plotted there will automatically have a z-coordinate equal to 5. Thus, for example, the projection of (2, 2, 2) should be plotted as the point (2.75, 3.125).

A chart of projected coordinates

For each of the viewpoints listed previously, the chart here gives the x- and y-coordinates of the projected vertices and a diagram of the projected cube, showing the visible edges with solid lines and the hidden edges with dotted lines. The axes shown are the x- and y-axes within the screen, that is, on the plane $z = 5$. The chart shows what happens to each of the vertices of the cube. For instance, the first example shows that the point (0, 0, 0) is projected to the point (2, 2.5).

viewpoint: (4, 5, 10)

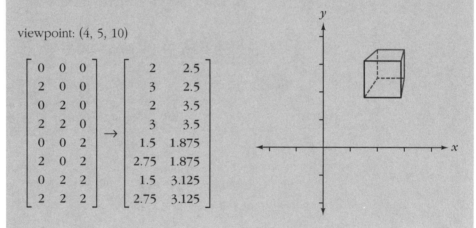

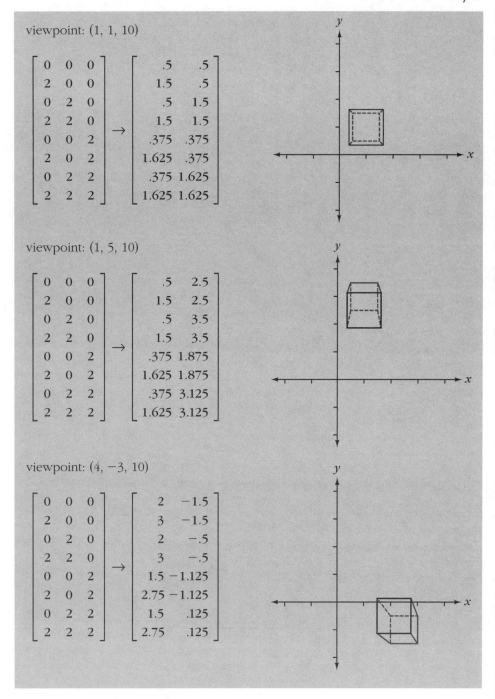

Day 28

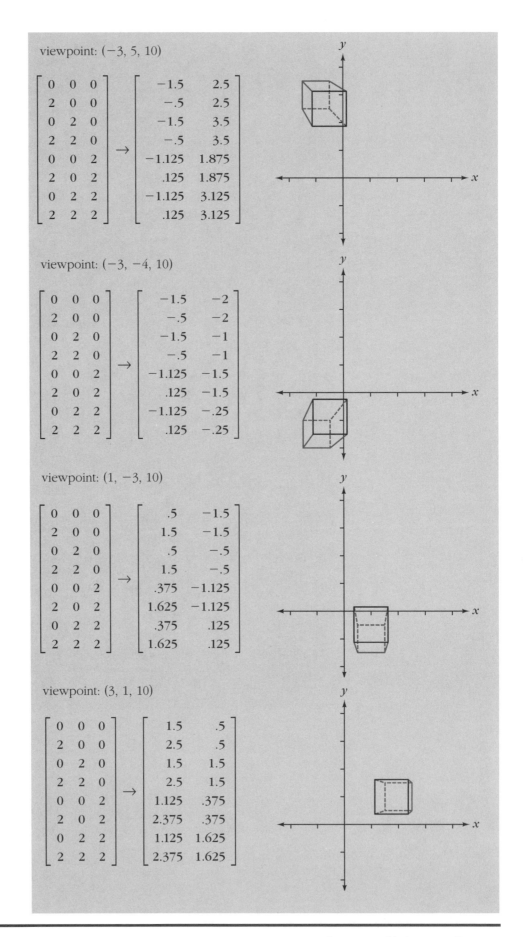

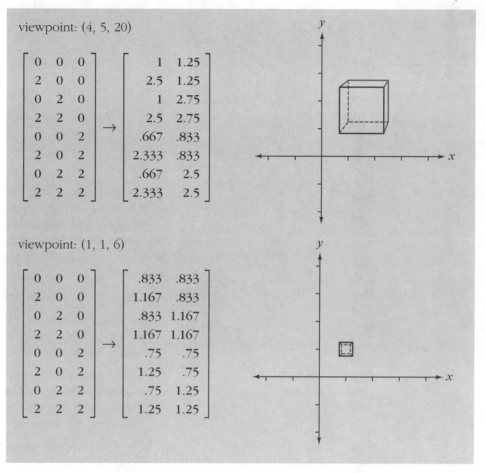

Homework 28: An Animated Outline

In this homework, students complete the outline of their program for *POW 5: An Animated POW*.

"The activity in As the Cube Turns *called* Find Those Corners! *was a pivotal activity for my classes. Most of the students went into the activity thinking that the projections would all look the same. They also wondered why I was giving each group a different perspective to consider. When they finished the activity and were able to compare the different groups' posters, they were amazed. This really gave them a sense of how important the perspective was, and a number of them made a connection to how this would look in their programs and how they would have to use the formulas that they had created."*

IMP teacher Michelle Novotny

Day 28

Find Those Corners!

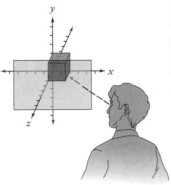

Let's fix a 2-by-2-by-2 cube in a three-dimensional coordinate system. For convenience, we'll place it snugly in the corner where the coordinate planes meet, so that three of the cube's faces are against the coordinate planes, with one vertex of the cube at (0, 0, 0) and the diagonally opposite vertex at (2, 2, 2).

1. Find the coordinates of the other six corners.

2. Imagine that the plane $z = 5$ is a screen. Your teacher will give you a viewpoint. Write down the coordinates of this viewpoint.

Using the plane $z = 5$ as the screen and the viewpoint you are given, determine the coordinates where each of the vertices of the cube will be projected on the screen.

3. Plot the projections of the vertices you found in Question 2 on a large piece of graph paper, thinking of the piece of graph paper as the plane $z = 5$. Put an appropriate scale on the axes.

Label the projections for the vertices (2, 2, 2), (0, 0, 0), and so on, and connect them to draw the cube. Use solid lines for the edges that are visible and dotted lines for the edges that are hidden by the rest of the cube.

Comment: If your drawing doesn't look something like a cube, you will need to revise your work.

An Animated Outline

In this assignment, you and your partner will complete the outline for your animation project, *POW 5: An Animated POW*. You will turn it in tomorrow, but you should keep a copy so you can continue your work.

The next stage after this assignment will be to write the program itself, beginning with a plain-language program. Remember to keep written copies of all your work on the program. This will reduce the risk of losing valuable work. It may also be easier to find errors in your written copy than on the calculator.

Plan to have your program entered well before your presentation day. Usually, "debugging" your program takes more time than writing the first draft.

DAY 29 Finding More Corners

Students continue to find the corners.

Mathematical Topics

- Finding projections

Outline of the Day

In Class

1. Collect *Homework 28: An Animated Outline*
2. Complete *Find Those Corners!*
   - Discussion will take place on Day 30

At Home

Homework 29: Mirrors in Space

1. Collection of *Homework 28: An Animated Outline*

Collect students' outlines and check them for problems. You may want to remind students to keep written copies of their programs and to leave enough time for debugging.

2. Continued Work on *Find Those Corners!*

Let students continue working on yesterday's activity, *Find Those Corners!* They should complete their posters by the end of class and display them. Tell them that their posters should not list the viewpoints they used, because everyone will be trying to match viewpoints to the diagrams on the posters, based on specific details of the diagrams.

Use your observations of groups to guide you as to whether a presentation on the numerical details is needed. If possible, the focus of the discussion tomorrow should be on the process of matching diagrams and viewpoints.

Day 29

Homework 29: Mirrors in Space

This homework is a three-dimensional analog of *Homework 24: Flipping Points*.

Ryane Snow and Dennis Cavaillé discuss what questions they might pose when their students work on "Find Those Corners!"

Mirrors in Space

You've looked at coordinate and matrix representations in two dimensions for each of the three basic types of isometries: *translations, rotations,* and *reflections*.

In this assignment, you will explore reflections in 3-space. In this setting, we use a **plane of reflection,** which is analogous to a *line* of reflection in two dimensions. The reflection of a point P through a plane m is the point Q that makes m the perpendicular bisector of $\overline{PQ}$. In other words, $\overline{PQ}$ is perpendicular to m, and m intersects the segment at the midpoint of the segment. The plane m will be a plane of symmetry between any set of points and the set of their reflections through m.

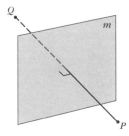

1. a. Begin with the point $(4, 6, 2)$, and imagine reflecting that point using the yz-plane as the plane of reflection. (The yz-plane is the same as the plane $x = 0$.) Determine where the point $(4, 6, 2)$ should end up.

b. Now generalize your work from Question 1a, using the point (x, y, z). That is, find the coordinates of the point you would get if you reflected the point (x, y, z) through the yz-plane.

2. Find a matrix A that will do the work of Question 1b for you. That is, find a matrix A so that the matrix product $[x \ y \ z] A$ gives the coordinates of the reflection of the point (x, y, z) in the yz-plane.

DAY 30

Finishing the Corner

Mathematical Topics

• Expressing reflections in three dimensions using matrices
• Working with spatial reasoning
• Continuing work with projections

Students reverse the process and find the viewpoint by looking at the projection.

Outline of the Day

In Class

1. Discuss *Homework 29: Mirrors in Space*
2. Discuss *Find Those Corners!*
   • Have students try to determine which viewpoint matches which projection, and have them describe their reasoning
3. Develop and post a procedure for finding the coordinates of a projected point

At Home

Homework 30: Where Are We Now?

1. Discussion of *Homework 29: Mirrors in Space*

Have students compare answers on the homework within their groups. For variety, you may want to choose a suit at random and have students with that suit card from one or more groups report on each question.

The use of the specific point (4, 6, 2) in Question 1a should be helpful for getting students to see what is happening with the coordinates. They should see that in general, the point (x, y, z) is reflected to $(-x, y, z)$.

Question 2 here is similar to Question 3 of *Homework 24: Flipping Points*. If students need a hint, you can suggest that they write out the question in matrix form, something like the equation on the next page.

Day 30

$$[x \ y \ z] \begin{bmatrix} ? & ? & ? \\ ? & ? & ? \\ ? & ? & ? \end{bmatrix} = [-x \ y \ z]$$

This should lead them to see that the reflection matrix should be

$$\begin{bmatrix} -1 & 0 & 0 \\ 0 & 1 & 0 \\ 0 & 0 & 1 \end{bmatrix}$$

2. Discussion of *Find Those Corners!*

If groups haven't already done so, they should put up their posters. Give everyone 5 to 10 minutes to try to match the pictures with the posted list of viewpoints.

The discussion should focus on how students did their matching. (We are assuming that they didn't actually go through the projection for every viewpoint.) You can let volunteers point to a specific poster (other than their own) and explain how they figured out which viewpoint it represented.

For example, the first diagram in the chart on Day 28 shows that the viewpoint is above and to the right of the cube, because the front, right, and top faces of the cube are visible. The only viewpoints in the list that fit this condition are (4, 5, 10) and (4, 5, 20). Of these two, the first gives a perspective more from the side, because it is closer, while (4, 5, 20) shows more of a head-on view of the cube, because it is farther away.

Similarly, from the viewpoint (1, 5, 10), one should see only the front and top faces, and there is only one diagram that fits this condition.

You may not need to go over the computational details for any of the examples, depending on how comfortable the groups seemed to be with the computations. Following the matching activity, students will develop a general outline for how to do this, and you can review the computational details as part of that work.

One detail you may wish to bring out now is that although the cube has eight vertices, there are only two r-values in the computations. That is, the vertices on the front face all give one value for the ratio $\frac{\text{distance to screen}}{\text{distance to viewpoint}}$, while the vertices on the back face all give a second value for this ratio.

"Are there always only two r values?"

Ask students if this is always the case for every choice of screen, viewpoint, and position of the cube. They should see that it holds true only when the cube is positioned so that it has two of its faces parallel to the screen.

3. How to Project

Now, ask students to work in groups to write out algebraic steps for projecting a point onto the screen. For uniformity of notation, you may want to suggest that they represent the point being projected as (x_1, y_1, z_1), the viewpoint as (x_2, y_2, z_2), and the screen as the plane $z = k$.

When students seem ready, bring them together to develop a class outline for this. Tell them that they will use this outline for writing the final program on Day 33.

Here is a sample outline. (*Note:* Again, we are thinking of the screen as being part of the way from the vertex to the viewpoint, rather than part of the way from the viewpoint to the vertex.)

Step 1. Find the ratio r: Find the difference in z-coordinates between the projected point and the screen, find the difference in z-coordinates between the projected point and the viewpoint, and find the ratio of these two z-distances. This gives

$$r = \frac{k - z_1}{z_2 - z_1}$$

Step 2. Find the x-coordinate of the projection: Find the difference in x-coordinates between the projected point and the viewpoint, multiply this x-distance by r, and add the result to the x-coordinate of the point. This gives

$$x_{\text{projection}} = x_1 + r(x_2 - x_1)$$

Step 3. Find the y-coordinate of the projection: Find the difference in y-coordinates between the projected point and the viewpoint, multiply this y-distance by r, and add the result to the y-coordinate of the point. This gives

$$y_{\text{projection}} = y_1 + r(y_2 - y_1)$$

Step 4. Plot the point: Plot the point using the x- and y-coordinates found in steps 2 and 3.

Post this description prominently. You can now check off item 4 from the unit plan from Day 1: "Create a two-dimensional drawing of a three-dimensional object."

On Days 31 and 32, students will look at item 5: "Change the position of an object located in a three-dimensional coordinate system." (As noted on Day 11, this will involve looking only at rotations.) In *Homework 32: The Turning Cube Outline,* they will look at the task of putting all the pieces together.

Homework 30: Where Are We Now?

This assignment gives students another chance to look at where they are in the unit.

Day 30

Where Are We Now?

This is a good time to reflect on the unit. Your tasks in this assignment are to describe the unit goal, to summarize what has happened in the unit so far, and to indicate what you think remains to be done.

1. Describe what your expectations were early in the unit. That is, what did you think the unit would be about? What mathematics did you think you would be learning?

2. What have turned out to be the key mathematical ideas of the unit? How do they relate to the unit problem?

3. What ideas or procedures do you think you still need to learn in order to solve the unit problem?

Day 31

DAYS 31-33
Rotating in Three Dimensions

This page in the student book introduces Days 31 through 33.

Joel Picazo uses coordinates and matrices to rotate in three dimensions.

You are finally ready for the cube! You need only combine what you know about rotations in two dimensions with what you know in general about three dimensions to create matrices for rotating in three dimensions. Then you will be turning the cube!

DAY 31

Students learn about rotations in three dimensions.

Rotating Around It

Mathematical Topics

- Reviewing the main ideas of the unit
- Visualizing rotations about an axis in three dimensions
- Expressing rotations in three dimensions in terms of coordinates

Special Materials Needed

- A large Styrofoam cube (or similar object) and two sticks for a classroom demonstration
- A cube for each student

Outline of the Day

In Class

1. Discuss *Homework 30: Where Are We Now?*
   - Have students share their perceptions from the beginning of the unit, their views about the key mathematical ideas, and their thoughts about what's still needed
2. Introduce rotations around the z-axis
   - Have students review rotations in two dimensions
   - Use a physical model to help students picture what it means to rotate the cube around the z-axis
3. *Follow That Point!*
   - Students rotate a pair of points about the z-axis
4. Discuss *Follow That Point!*
   - Emphasize that rotation around the z-axis is like rotation in the xy-plane, with the z-coordinate kept fixed

At Home

Homework 31: *One Turn of a Cube*

1. Discussion of *Homework 30: Where Are We Now?*

Ask various diamond card students to contribute what they thought the unit would entail and what mathematics they thought they would be learning.

Interactive Mathematics Program   As the Cube Turns 223

Then ask other diamond card students to explain the important ideas of the unit so far and how they relate to the unit problem.

Next, ask other diamond card students to contribute thoughts on what they have to do next, in order to get an idea of where the class stands in the process of solving the unit problem. This homework is a good one for you to collect to get a sense of how well students understand the unit so far.

2. Rotating Around the z-axis

Tell students that they are now going to look at the last piece of the grand plan, item 5, as modified on Day 11:

> **5. Change the position of an object located in a three-dimensional coordinate system.**
>
> - **Rotate the object.**

- *Review from two dimensions*

    Ask for one or more volunteers to review what the class knows about rotation in two dimensions. Elicit statements in terms of both coordinates and matrices.

    In terms of coordinates, you should have something like this posted from Day 15:

    > **If a point (x, y) is rotated counterclockwise through an angle ϕ, then the coordinates (x', y') of its new location are given by the equations**
    > $$x' = x \cos \phi - y \sin \phi$$
    > $$y' = x \sin \phi + y \cos \phi$$

    In terms of matrices, you should have something like this posted from Day 17:

    > **To rotate a matrix of two-dimensional points counterclockwise around the origin by an angle ϕ, multiply that matrix on the right by the matrix**
    > $$\begin{bmatrix} \cos \phi & \sin \phi \\ -\sin \phi & \cos \phi \end{bmatrix}$$

- *What is a rotation in three dimensions?*

    The next task is to be sure that students have a visual picture of what is meant by rotation in three dimensions about an axis. As described and diagrammed on Day 1, we are imagining the blackboard as the xy-plane, with the z-axis coming "forward." Place a cube "in space" and ask students for the coordinates of its vertices. It may help to use a stick attached to a Styrofoam cube to demonstrate the process, as described on Day 1.

Day 31

Then ask what would happen to the coordinates if the object were rotated around the z-axis. Each student should have a small cube to work with to get a sense of what is happening.

Let various students come up to show the rotation. They should see that the z-coordinate of each point stays the same, because each point is traveling within a plane parallel to the xy-plane. Thus, students can essentially ignore the z-coordinate and treat the motion of each point within that plane. The next activity will make this more concrete by looking at a numerical example.

3. *Follow That Point!*

Let groups begin work on *Follow That Point!* If they are struggling with Question 1, you can suggest that they imagine the graph paper as if it were in the plane $z = -5$. They should think of a point as rotating around the origin on the piece of paper.

4. Discussion of *Follow That Point!*

Ask several club card students to present their answers and reasoning.

Students should see that each of the two points under consideration is being rotated in a plane parallel to the xy-plane. That is, the point $(2, 3, -5)$ is rotating within the plane $z = -5$, and the point $(5, 2, 6)$ is rotating within the plane $z = 6$. Thus, students can work with the x- and y-coordinates as if the points were in the xy-plane itself.

If $(2, 3)$ were rotated counterclockwise 20° around the origin, it would move to $(2 \cos 20° - 3 \sin 20°, 3 \cos 20° + 2 \sin 20°)$. Therefore, when $(2, 3, -5)$ is rotated counterclockwise 20° around $(0, 0, -5)$ (within the plane $z = -5$), it moves it to $(2 \cos 20° - 3 \sin 20°, 3 \cos 20° + 2 \sin 20°, -5)$, which is approximately $(0.85, 3.50, -5)$.

> *Note:* If students go back to basics, they may express their answer to Question 1 as something like this:
>
> $$\left(\sqrt{13} \cos\left[\tan^{-1}\left(\tfrac{3}{2}\right) + 20°\right], \sqrt{13} \sin\left[\tan^{-1}\left(\tfrac{3}{2}\right) + 20°\right], -5\right)$$
>
> If any students do this, you can use it as an occasion to review why this approach doesn't necessarily work in all quadrants, bringing out that this method would give the same result if they started with $(-2, -3, 5)$ instead of $(2, 3, 5)$. You might also review the way the rotation formula was developed, using the sine-of-a-sum and cosine-of-a-sum formulas.

Day 31

- *Question 2*

    On Question 2, the main goal is to bring out that the result of this rotation can be described as the segment whose endpoints are the results of rotating the original endpoints.

    Students should be able to find the result of rotating $(5, 2, 6)$ just as they did with $(2, 3, -5)$. A $20°$ counterclockwise rotation about the z-axis takes the point $(5, 2, 6)$ to the point $(5 \cos 20° - 2 \sin 20°, 2 \cos 20° + 5 \sin 20°, 6)$, which is approximately $(4.01, 3.59, 6)$. Thus, the segment connecting $(2, 3, -5)$ and $(5, 2, 6)$ ends up after rotation as the segment connecting $(0.85, 3.50, -5)$ and $(4.01, 3.59, 6)$.

Homework 31: One Turn of a Cube

> This assignment continues the work with rotations in three dimensions.

Follow That Point!

1. Consider the point $(2, 3, -5)$, and imagine that this point is rotated 20° counterclockwise around the z-axis. Where does this point end up?

2. Now suppose you have a segment connecting $(2, 3, -5)$ and $(5, 2, 6)$, and you rotate that segment 20° counterclockwise around the z-axis. What is the result of this rotation?

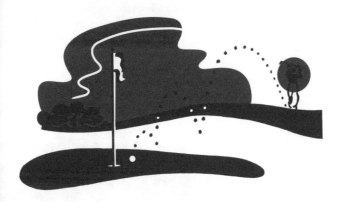

One Turn of a Cube

Imagine placing a cube in 3-space. For simplicity, choose a placement for which the faces of the cube are parallel to the coordinate planes.

1. Write down the coordinates in 3-space for the eight vertices of your cube.

2. Imagine that the cube has been rotated 30° counterclockwise around the z-axis. Choose one face of the cube, and find the new coordinates of the vertices on that face. (Round off the coordinates to the nearest tenth.)

3. Make two sketches, one showing your cube in its original position in 3-space and another showing its position after rotation.

DAY 32

Students use matrices to express rotations in 3-space.

Rotating Matrices

Mathematical Topics

- Expressing rotations in three dimensions using matrices

Outline of the Day

In Class

1. Discuss *Homework 31: One Turn of a Cube*
   - Emphasize that the z-coordinates stay fixed
2. *Rotation Matrix in Three Dimensions*
   - Students find matrices for rotations around each of the three axes
3. Discuss *Rotation Matrix in Three Dimensions*
   - Have students use their matrix from Question 1 to check last night's homework

At Home

Homework 32: The Turning Cube Outline

---

1. Discussion of *Homework 31: One Turn of a Cube*

Have students confer in their groups. It is important for them to see that the z-coordinate stays fixed and that the rotation is like a two-dimensional rotation involving only the x- and y-coordinates. Each point moves around in a two-dimensional plane of the form $z = k$.

To illustrate the algebra of the rotation, suppose one of the vertices of the cube is at $(3, 5, 7)$. The rotated point will still have a z-coordinate equal to 7. The new x- and y-coordinates, which we label here as x' and y', are found by the usual formulas:

$$x' = x \cos 30° - y \sin 30°$$
$$y' = x \sin 30° + y \cos 30°$$

In other words, the rotated point has these new coordinates:

$$(3 \cos 30° - 5 \sin 30°, 3 \sin 30° + 5 \cos 30°, 7)$$

Interactive Mathematics Program

As the Cube Turns 229

Day 32

which is approximately (0.10, 5.83, 7). (Students should verify that this seems reasonable on the graph.)

Note: If students placed their cubes so that some vertices were on the z-axis, they should have seen that these vertices did not move when the cube was rotated around the z-axis.

2. Rotation Matrix in Three Dimensions

This activity should be fairly straightforward in light of the homework discussion and all the work students have done so far with matrices.

3. Discussion of *Rotation Matrix in Three Dimensions*

Have a spade card student report results. Students should be able to figure out that the rotation matrix B should be

$$B = \begin{bmatrix} \cos 30° & \sin 30° & 0 \\ -\sin 30° & \cos 30° & 0 \\ 0 & 0 & 1 \end{bmatrix}$$

"Can you verify that the rotation matrix works?"

Ask students to verify that this works, not just for a single point but also for a *matrix* of three-dimensional points. That is, they should take their own vertex matrices from last night's homework, multiply them by this matrix B, and see that they do get the coordinates of the rotated vertices. Even if students did not develop the matrix B on their own, they should see that it does the job.

Before looking at rotations around other axes, ask students to generalize Question 1 for an arbitrary angle of rotation. You should be able to post a general statement something like this:

> **To rotate a matrix of three-dimensional points counterclockwise around the z-axis by an angle ϕ, multiply that matrix on the right by the matrix**
>
> $$\begin{bmatrix} \cos \phi & \sin \phi & 0 \\ -\sin \phi & \cos \phi & 0 \\ 0 & 0 & 1 \end{bmatrix}$$

- Question 2

For the issue of rotation around other axes, students may see that they simply need to rearrange the entries of the rotation matrix. For example, for rotation around the x-axis, the matrix looks like this:

230 *As the Cube Turns*

Interactive Mathematics Program

$$\begin{bmatrix} 1 & 0 & 0 \\ 0 & \cos\phi & \sin\phi \\ 0 & -\sin\phi & \cos\phi \end{bmatrix}$$

For rotation around the *y*-axis, one multiplies the matrix on the right by

$$\begin{bmatrix} \cos\phi & 0 & \sin\phi \\ 0 & 1 & 0 \\ -\sin\phi & 0 & \cos\phi \end{bmatrix}$$

Homework 32: The Turning Cube Outline

This assignment is students' final preparation for solving the unit problem. You may want to review the generic animation outline developed on Day 9:

Setup program

Set the initial coordinates

Start the loop
- Clear the screen
- Draw the next figure
- Delay
- Change the coordinates

End the loop

You may want to remind students that along with setting the initial conditions, they also need to set the "change matrix," which would be the matrix of rotation.

"Many students went beyond what I expected in their cube project and incorporated their new programming skills with anything they could get from other sources."

IMP teacher Moe Burkhart

Rotation Matrix in Three Dimensions

You have seen that, if a point (x, y, z) is rotated counterclockwise around the z-axis, through an angle of $30°$, its new coordinates are

$(x \cos 30° - y \sin 30°, x \sin 30° + y \cos 30°, z)$

1. Find a rotation matrix for this transformation. That is, find a matrix B so that

$[x \; y \; z] \, B =$
$\quad [x \cos 30° - y \sin 30° \quad x \sin 30° + y \cos 30° \quad z]$

2. What should the rotation matrix be if the rotation is around the x-axis? The y-axis?

The Turning Cube Outline

Writing animation programs is a very complicated task, and the unit problem is no exception. You have had to learn many ideas about mathematics and programming to develop a program for the turning cube, and you now have all the necessary pieces.

Most programmers find it helpful to write an outline for a program before developing the code. Your task in this assignment is to look over your work for the unit, and develop an outline for a program to turn the cube. You need not give line-by-line details. Instead, give the general structure of how the program should be organized.

As the Cube Turns

Mathematical Topics

• Solving the unit problem

Outline of the Day

In Class

1. Discuss *Homework 32: The Turning Cube Outline*
   • Generate a class outline
2. Concluding *As the Cube Turns*
   • Select one of the teacher options for concluding the unit
   • Implement your decision

At Home

Homework 33: Beginning Portfolio Selection

Discuss With Your Colleagues

But We've Come So Far...!

Students have spent much time and energy working up to writing a program to turn the cube. Yet you may decide they would be better off not spending the time entering a program into their calculators. Such an exercise may bring more frustration than enlightenment. What might be the effect on your students of not letting them write their own programs to turn the cube?

1. Discussion of *Homework 32: The Turning Cube Outline*

Let groups work together, sharing ideas for a few minutes, and then bring the class together to develop a common outline. As a first stage, you might get something like the outline on the next page.

Setup program

Set the coordinates for the viewpoint and for the initial vertices of the cube, set the equation for the screen, and set the rotation matrix

Start the loop
- Clear the screen
- Find the coordinates for the projections on the screen of the vertices of the cube
- Draw the projected cube by connecting certain pairs of vertices
- Delay
- Change the coordinates of the vertices of the cube using a three-dimensional rotation matrix

End the loop

The most subtle element of this outline is the fact that the vertices of the *actual* cube must be rotated, and not merely the vertices of the *projected* cube. You may want to go back to students' work on *Cube on a Screen* (Day 27) to help them clarify why this is necessary.

You may also want to remind students that on Day 30, they developed fairly detailed instructions for the step "Find the coordinates for the projections on the screen of the vertices of the cube."

2. Concluding *As the Cube Turns*

There are at least three options available to you for concluding the unit:

- You can let students work in groups to write a program to turn the cube.
- You can give students the TI-82 program TURNCUBE (see Appendix B), and have them explain how it works.
- You can have students explain TURNCUBE and then work on developing an improved version of the program.

Rather than put one of these options in the student materials, we include suggestions here for each option. Although this concluding task of the unit can be completed in one day (today), you may decide that it is worthwhile to allow a second day for students to do this work.

Your choice of option should depend on how well your students have understood the mechanics of programming and on your judgment about whether having them develop the details of the program on their own will enhance their understanding or obscure it. You may want to give students a voice in this decision.

IMP field-test teachers have used all three options. Even when they have used the second or third option, in which students do not write the program to turn the cube, students have felt it was a satisfactory conclusion to the unit.

- *Implementing your decision*

    Here are some ideas for each of the options.

    If students write the program: If you want to have students write their own program in groups, be sure to have them write it on paper before putting it into the calculator. If a link is available from computers to calculators (so that programs can be written on a computer and then transferred to the calculator), we strongly recommend that students use it, for these reasons:

    - Students can store their programs overnight in the computer so they won't be lost.
    - Students can enter and debug their programs on the computer, where editing is easier.

    A suggestion you might make: "What are the numerical ratios for the front and back faces?"

    One suggestion you might make that will simplify the programming is for students to find the two ratios for the front and back faces of the cube numerically, and simply use these numbers in the program, rather than compute them as part of the program. (We assume that students will place the cube so that two of its faces are parallel to the screen.)

    If students focus on explaining the program: If you want to focus on having students simply explain the program TURNCUBE, here are some specific tasks that you can give them:

    - Find the initial position of the cube.
    - Find the plane of the screen.
    - Find the angle for each rotation.
    - Explain the roles of specific variables.

    If students develop an improved version of the program: You may want to discuss as a class how the program might be improved. For instance, it might provide for user input on the choice of viewpoint, screen, and position of the cube.

Homework 33: Beginning Portfolio Selection

> Tonight's homework assignment gets students started compiling their portfolios for this unit, asking them to summarize how they learned about each item on their outline.

Beginning Portfolio Selection

Now that you have turned the cube, you can understand the different items on the outline you saw at the beginning of the unit. That outline probably looked something like this.

1. Draw a picture on the graphing calculator.
2. Create the appearance of motion.
3. Change the position of an object located in a two-dimensional coordinate system.
4. Create a two-dimensional drawing of a three-dimensional object.
5. Change the position of an object located in a three-dimensional coordinate system.

Choose three of the items from this outline. For each item, choose one assignment that was important in developing that item. Explain how that assignment helped you understand the given item and how that item fit into the overall development of the unit.

You will work with the remaining two items in *Homework 35: Continued Portfolio Selection*.

Day 34

An Animated POW

This page in the student book introduces Days 34 through 36.

Leah Allen and Lou Allen Wheeler work together to create their own animation to present to their classmates.

Now that you have turned the cube, you are ready to create an interesting animation of your own. Over the next several days, you will put together your work on *POW 5: An Animated POW,* present it, and see the animations your classmates have created.

DAY 34

Completely Animated

Mathematical Topics

• Designing and programming animation

Students complete work on their projects.

Special Materials Needed

- A calculator manual for each pair of students
- (Optional) A video on the making of animation features

Outline of the Day

In Class

1. Discuss *Homework 33: Beginning Portfolio Selection*
   • Have a few students share ideas
2. Have pairs complete work on *POW 5: An Animated POW*
3. (Optional) Show an animation video

At Home

Homework 34: "An Animated POW" Write-up

1. Discussion of *Homework 33: Beginning Portfolio Selection*

You can have a couple of volunteers read their descriptions of how each item from the outline fit into the development of the unit. You may also want to discuss what makes a good description.

Also have some students state their choices of assignments and give their explanations of how these assignments helped them understand the ideas involved.

Note: Because students will be working further with the unit outline from last night's homework (see *Homework 35: Continued Portfolio Selection*), we suggest that you have them identify clearly which assignment they selected goes with which item in the outline. Although this may seem obvious to them now, it may not be so clear in a couple of days.

Day 34

2. Completion of *POW 5: An Animated POW*

The rest of today is allotted for students to complete work on *POW 5: An Animated POW* (unless you decide to show an animation video—see the next section). Days 35 and 36 are scheduled for POW presentations.

You probably can accommodate about eight presentations per day, and the allocation of two days is based on a class with about 32 students (that is, 16 pairs).

You may want to have pairs draw cards to determine which pairs will present on which day. (For instance, have pairs drawing clubs and spades present one day and pairs drawing diamonds and hearts present the next day.)

Encourage students to share ideas with all of their classmates, not just with their partners, as they work on this project. All of them are learning not only about the mathematics involved but also about the capabilities of the graphing calculator. (You can have them include you as well as their classmates in this sharing.)

Comment: It might be worthwhile to try making a video of the students' presentations of these projects. You can show the video to other mathematics classes, to colleagues, and to administrators, and it can also be used with parents at an open house.

- *Peer reviews of presentations*

    You may want to have students write reviews of one another's POW presentations. If so, remind students of the grading criteria for the POW, as discussed on Day 20. We recommend that you have students brainstorm and create a rubric for evaluating one another's work, based on those criteria. (You may want to let them use different criteria from those you will use.) For simplicity, you might want to create a form that students can use in writing these reviews.

3. Optional: Animation Video

As noted on Day 1, there are videos on the making of animation features, such as *The Making of Toy Story*, that your students may appreciate after completing this unit. A good video can help students see that the mathematics they have learned really is used by professional animators. It can also help them see how complicated it is to make a feature animation.

Homework 34: "An Animated POW" Write-up

Day 34

> Students' only homework tonight is to complete their write-ups for their work on *POW 5: An Animated POW*.

Tell students that write-ups for *POW 5: An Animated POW* will be part of their portfolios for this unit.

If you have told students which day their presentation will be, you may want to suggest that groups presenting on Day 36 do this assignment tomorrow night and do *Homework 35: Continued Portfolio Selection* tonight.

Katie Davidow and Talia Rubinow work together to create their own animation to present to their classmates.

"An Animated POW" Write-up

You and your partner should now have almost completed your work on *POW 5: An Animated POW*.

Your homework tonight is to complete your work on this POW. This has several parts.

1. Complete the program.
2. Prepare your written copy of the program.
3. Develop a presentation for the class, lasting three to four minutes. Your presentation should include
   - A demonstration of the program
   - A description of one interesting feature of the program

DAY 35

POW 5 Presentations

Student pairs make presentations on the animation programs.

Mathematical Topics

• Giving and reviewing oral presentations

Outline of the Day

In Class
Presentations of *POW 5: An Animated POW*
• (Optional) Have students work within groups to review each presentation

At Home
Homework 35: Continued Portfolio Selection

Presentations of *POW 5: An Animated POW*

Today and tomorrow are set aside for presentations of *POW 5: An Animated POW*.

If students are evaluating each other's presentations (see the subsection "Peer reviews of presentations" on Day 34), we suggest that you have them do this immediately after each presentation. Students can work together within their groups to write some comments on the program and the presentation. You can pass these reviews on to the presenters after you have read them.

Be sure to allow groups a reasonable amount of time to discuss and evaluate the quality of each presentation and to write their reports before moving on to the next presentation. You can use this time to give each group your own grade. You will probably want to collect the group reviews on a given presentation before beginning the next one.

Homework 35: Continued Portfolio Selection

Tonight's homework continues work in compiling student portfolios.

Continued Portfolio Selection

In this assignment, you continue the work on your portfolio that you began in *Homework 33: Beginning Portfolio Selection*. In that assignment, you looked at the different stages in the outline you saw at the beginning of the unit:

1. Draw a picture on the graphing calculator.

2. Create the appearance of motion.

3. Change the position of an object located in a two-dimensional coordinate system.

4. Create a two-dimensional drawing of a three-dimensional object.

5. Change the position of an object located in a three-dimensional coordinate system.

In the earlier assignment, you selected three of the items from this outline. In tonight's assignment, you should work with the remaining two items. As before, for each of these items, choose one assignment that was important in developing that item, explain how the assignment helped you understand the item, and explain how the item fit into the overall development of the unit.

DAY 36 Finishing Presentations

More students make presentations on their animation programs.

Mathematical Topics

• Giving and reviewing oral presentations

Outline of the Day

In Class
1. Discuss *Homework 35: Continued Portfolio Selection*
2. Continue presentations of *POW 5: An Animated POW*
   • (Optional) Have students work within groups to review each presentation

At Home
Homework 36: "As the Cube Turns" Portfolio

1. Discussion of *Homework 35: Continued Portfolio Selection*

Because different students likely chose different items from the outline for *Homework 33: Beginning Portfolio Selection*, you may have covered all five items in the Day 34 discussion and not need to have any homework discussion today. (*Note:* As with *Homework 33: Beginning Portfolio Selection,* we suggest that you have students identify clearly which assignment they selected goes with which item in the outline.)

2. Continued Presentations on *POW 5: An Animated POW*

Finish the group presentations as described on Day 35.

Interactive Mathematics Program   As the Cube Turns

Day 36

Homework 36: "As the Cube Turns" Portfolio

For homework tonight, students will complete their unit portfolios. Be sure that students bring back the portfolio tomorrow with the cover letter as the first item. They should also bring to class any other work that they think will be of help on the unit assessments. The remainder of their work can be kept at home.

"My students enjoyed working on their animation projects. For 2½ days, they were totally engrossed with creating their animations, researching commands in the TI-83 manual, working with their partners, and getting their programs to work. They even asked permission to take home the calculators to finish their projects. The class developed a rubric to evaluate the animation projects and did a good job grading each other's presentations."

IMP teacher Wendy Tokumine

As the Cube Turns Portfolio

Now that *As the Cube Turns* is completed, it is time to put together your portfolio for the unit. Compiling this portfolio has three parts.

- Writing a cover letter in which you summarize the unit
- Choosing papers to include from your work in this unit
- Discussing your personal growth during the unit

Cover Letter for *As the Cube Turns*

Look back over *As the Cube Turns* and describe the central problem of the unit and the main mathematical ideas. This description should give an overview of how the key ideas were developed in this unit and how they were used to solve the central problem.

Continued on next page

Selecting Papers from *As the Cube Turns*

Your portfolio for *As the Cube Turns* should contain these items.

- *Homework 17: How Did We Get Here?*
- *Homework 27: Spiders and Cubes*
- *Homework 33: Beginning Portfolio Selection* and *Homework 35: Continued Portfolio Selection*

  Include your own work on the activities from the unit that you selected in these assignments.

- A Problem of the Week

  Select one of the first two POWs you completed during this unit (*"A Sticky Gum Problem" Revisited* or *A Wider Windshield Wiper, Please*).

- Your work on *POW 5: An Animated Project*

  Include both *Homework 28: An Animated Outline* and *Homework 34: "An Animated POW" Write-up*.

Personal Growth

Your cover letter for *As the Cube Turns* describes how the unit develops. As part of your portfolio, write about your own personal development during this unit. Because this is the first unit with a significant focus on programming, you may want to address this issue.

> *How have you grown in your understanding of the task of writing and interpreting programs?*

You should include here any other thoughts you might like to share with a reader of your portfolio.

DAY 37

Final Assessments

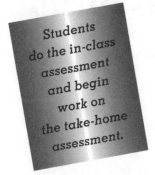

Students do the in-class assessment and begin work on the take-home assessment.

Special Materials Needed

- In-Class Assessment for "As the Cube Turns"
- Take-Home Assessment for "As the Cube Turns"

Outline of the Day

In Class
Introduce assessments
- Students do *In-Class Assessment for "As the Cube Turns"*
- Students begin *Take-Home Assessment for "As the Cube Turns"*

At Home
Students complete *Take-Home Assessment for "As the Cube Turns"*

End-of-Unit Assessments

The in-class assessment is intentionally quite short so that time pressures will not be a factor in students' ability to do well. The IMP *Teaching Handbook* contains general information about the purpose of end-of-unit assessments and how to use them.

Tell students that today they will get two tests—one that they will finish in class and one that they can start in class and finish at home. The take-home part should be handed in tomorrow.

Tell students that they are allowed to use graphing calculators, notes from previous work, and so forth when they do the assessments. (They will have to do without graphing calculators on the take-home portion unless they have their own.)

These assessments are provided separately in Appendix C for you to duplicate.

In-Class Assessment for *As the Cube Turns*

Here is a plain-language program for a graphing calculator:

Program: DRAW

Setup program

Let A be the matrix $\begin{bmatrix} 1 & 1 \\ 3 & 3 \\ 5 & 1 \end{bmatrix}$

Draw a line from (a_{11}, a_{12}) to (a_{21}, a_{22})

Draw a line from (a_{11}, a_{12}) to (a_{31}, a_{32})

Draw a line from (a_{31}, a_{32}) to (a_{21}, a_{22})

1. On a piece of graph paper, draw what the screen should look like after running this program using an appropriate viewing window. Show scales for the axes.

2. Use a loop and a matrix *B* to modify the program DRAW so the screen will look like this after running the new program.

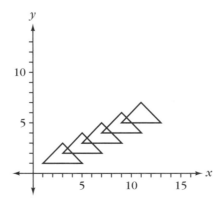

Continued on next page

3. Use a loop and a matrix C to modify the program DRAW so the screen will look like this after running the new program.

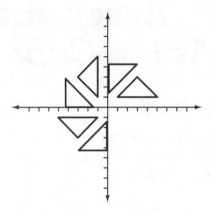

Note: The scales in this diagram are the same as in Question 2, although the viewing window is different. The numbers on the scales have been omitted to avoid cluttering the diagram.

Take-Home Assessment for *As the Cube Turns*

Part I: Where Are We?

This pair of equations will project a point (u, v, w) onto a screen parallel to the *xy*-plane so that the *x*- and *y*-coordinates of the projected point are a and b.

$$a = u + \left(\frac{6-w}{2-w}\right)(0-u)$$

$$b = v + \left(\frac{6-w}{2-w}\right)(1-v)$$

1. What are the coordinates of the viewpoint for the projection?

2. What is the equation of the screen?

3. Explain in detail why the equations given for a and b give the projection you describe.

Part II: How Does It Turn?

You have seen that when a point (x, y) is rotated counterclockwise around the origin through an angle ϕ, the new point has coordinates $x \cos \phi - y \sin \phi$ and $x \sin \phi + y \cos \phi$.

Explain how those two formulas were developed.

DAY 38

Summing Up

Students sum up what they learned in the unit.

Mathematical Topics

• Summarizing the unit

Outline of the Day

1. Discuss unit assessments
2. Sum up the unit

Note: The assessment discussion and unit summary are presented as if they take place on the day following the assessments, but you may prefer to delay this material until you have looked over students' work on the assessments. These discussion ideas are included here to remind you that you should allot some time for this type of discussion.

1. Discussion of Unit Assessments

You can have students volunteer to explain their work on each of the problems.

• *In-class assessment*

You can have volunteers present each question. Question 1 is a straightforward assessment of whether students understand the matrix notation. The screen might be shown like this:

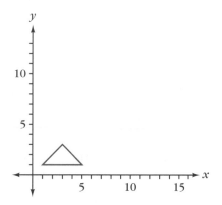

(Students may have used the first triangle in the diagram in Question 2 to confirm the correctness of their results.)

Interactive Mathematics Program   As the Cube Turns 255

For Question 2, the modified program will probably look something like this:

Setup program

Let A be the matrix $\begin{bmatrix} 1 & 1 \\ 3 & 3 \\ 5 & 1 \end{bmatrix}$

Let B be the matrix $\begin{bmatrix} 2 & 1 \\ 2 & 1 \\ 2 & 1 \end{bmatrix}$

For T from 1 to 5
- Draw a line from (a_{11}, a_{12}) to (a_{21}, a_{22})
- Draw a line from (a_{11}, a_{12}) to (a_{31}, a_{32})
- Draw a line from (a_{31}, a_{32}) to (a_{21}, a_{22})
- Replace matrix A by the sum A + B

End the T loop

Comment: Out of habit, some students might include "delay" and "clear screen" instructions in their programs. If so, bring out that the final screen shows all five triangles, so the "clear screen" instruction should not be included, and the "delay" instruction is not needed.

Question 3 is similar, with each triangle being a 45° counterclockwise rotation around the origin from the previous triangle. The plain-language program might look like this:

Setup program

Let A be the matrix $\begin{bmatrix} 1 & 1 \\ 3 & 3 \\ 5 & 1 \end{bmatrix}$

Let C be the matrix $\begin{bmatrix} \cos 45 & \sin 45 \\ -\sin 45 & \cos 45 \end{bmatrix}$

For T from 1 to 6
- Draw a line from (a_{11}, a_{12}) to (a_{21}, a_{22})
- Draw a line from (a_{11}, a_{12}) to (a_{31}, a_{32})
- Draw a line from (a_{31}, a_{32}) to (a_{21}, a_{22})
- Replace matrix A by the product A · C

End the T loop

Comment: Students might use numerical values for sin 45° and cos 45°, both of which are approximately .71.

- *Take-home assessment*

    For Part I, you might go straight to the explanation (Question 3), because that should also include the answers to Questions 1 and 2. The presenter should explain that $\frac{6-w}{2-w}$ gives the ratio between these two values:

    - The difference in z-coordinates between (u, v, w) and the screen
    - The difference in z-coordinates between (u, v, w) and the viewpoint

    The expression for the numerator shows that the screen has equation $z = 6$, and the expression for the denominator shows that the z-coordinate of the viewpoint is 2.

    The presenter should also explain that the expressions $0 - u$ and $1 - v$ give the x- and y-distances from (u, v, w) to the viewpoint, so the x- and y-coordinates of the viewpoint are 0 and 1, respectively. Thus, the viewpoint is at $(0, 1, 2)$.

    For Part II, you can have volunteers share ideas about the development of the rotation formulas. This will probably include the use of polar coordinates to describe the beginning and ending position of the point and the role of the sine-of-a-sum and cosine-of-a-sum formulas in expressing the result in terms of just $x, y,$ and ϕ.

2. Unit Summary

You can let students share their portfolio cover letters as a way to start a summary discussion of the unit.

Then let them brainstorm ideas of what they have learned in this unit. This is a good opportunity to review terminology and to place this unit in a broader mathematics context.

APPENDIX A

Supplemental Problems

This appendix contains a variety of activities that you can use to supplement the regular unit material. These activities fall roughly into two categories.

- Reinforcements, which are intended to increase students' understanding of and comfort with concepts, techniques, and methods that are discussed in class and that are central to the unit
- Extensions, which allow students to explore ideas beyond the basic unit and which sometimes deal with generalizations or abstractions of ideas that are part of the main unit

The supplemental activities are given here in the approximate sequence in which you might use them in the unit. In the student book, they are placed in the same order following the regular materials.

Here are some specific recommendations about how each activity might work within the unit. (For more ideas about the use of supplemental activities in the IMP curriculum, see the IMP *Teaching Handbook*.)

Loopy Arithmetic (extension)

The programming tasks in this problem ask students to make more sophisticated use of For/End loops than required by the unit problem. This assignment can be used once students have some basic experience with For/End loops, such as after Day 4.

Sum Tangents (reinforcement)

This activity follows smoothly from the development of the formulas for the sine and cosine of a sum on Days 13 through 15, and can be used anytime after the development of those formulas.

Moving to the Second Quadrant (reinforcement)

In *The Sine of a Sum*, students proved the formula for the sine of the sum of two acute angles. They were told that the formula they found works for all angles, but no proof was provided for other angles. In this activity, students prove the formula for the case in which one of the angles is in the second quadrant and the other is in the first quadrant. This activity can be done anytime after Day 15, when the formula for the sine of a sum is applied to find the cosine of a sum. *Adding 180°* and *Sums for All Quadrants* continue this work.

Appendix A

Adding 180° (reinforcement)

This activity is similar to Questions 1 and 2 of the supplemental activity *Moving to the Second Quadrant*. The formula that students are asked to develop in Question 1 is needed in the supplemental problem *Sums for All Quadrants*.

Sums for All Quadrants (extension)

In this activity, students prove the sine-of-a-sum formula for all angles, using a quadrant-by-quadrant analysis. The activity is a follow-up to the two preceding supplemental problems.

Bugs in Trees (extension)

This activity provides students with an application of matrices that is quite different from the one in this unit. The activity will connect somewhat back to their work with probability, and it can be used anytime after *More Memories of Matrices* is discussed on Day 17.

Half a Sine (reinforcement)

This activity is a natural follow-up to the work on *Homework 18: Doubles and Differences*.

The General Isometry (extension)

This activity provides a larger context for the work students are doing with translations, rotations, and reflections. It can be assigned after *Homework 24: Flipping Points*, which introduces reflections.

Perspective on Geometry (extension)

This activity is a change of pace and might be used after the activity *Cube on a Screen* (Day 27).

Let the Calculator Do It! (extension and reinforcement)

In this activity, students write a program for the calculator (or a computer) to find the projection of a point on the screen given a viewpoint. This assignment should follow the discussion of *Find Those Corners!* on Day 30.

Appendix A

Appendix

Supplemental Problems

This page in the student book introduces the supplemental problems.

The supplemental problems for *As the Cube Turns* continue the unit's areas of emphasis—programming, trigonometry, matrices, and transformational geometry—although other topics appear as well. Here are some examples.

- *Loopy Arithmetic* and *Let the Calculator Do It!* give additional work on writing programs.

- *Sum Tangents* and *Half a Sine* look at the development of other trigonometric formulas.

- *Bugs in Trees* shows a new application for matrices.

- *The General Isometry* is a challenging activity about combining isometries.

SUPPLEMENTAL PROBLEM

Loopy Arithmetic

The Basic For/End Loop

For/End loops can be used in programs to do a variety of repetitive tasks, including arithmetic.

By combining this type of loop with a display of results on the screen, you can get your graphing calculator to show you the work it's doing.

For example, the screen for this plain-language program will show the calculator counting from 20 to 100.

> For A from 20 to 100
> • Display the number A on the screen
> End the A loop

```
95
96
97
98
99
100
DONE
```

Well, actually, you won't see most of the counting happen, because the numbers will whiz by on your screen. But when the program is finished running, it will look something like the screen shown at the left.

Beyond Counting

Your graphing calculator can do more challenging tasks than counting, though. Here are two programs for you to write.

1. Write a program using a loop to compute factorials (without using an explicit factorial command). Your program should ask the user for an input and then give the factorial of that number.

   Important: The challenge of this program is writing it without using the calculator's factorial instruction.

Continued on next page

2. You may be familiar with the Fibonacci sequence of numbers, which begins like this:

$$1, 1, 2, 3, 5, 8, 13, 21, 34, \ldots$$

In this sequence, each term is obtained by adding the two previous terms. For example, the term 34 comes from the sum $13 + 21$.

Your task in this problem is to write a program using a loop to compute and display the first 40 Fibonacci numbers.

SUPPLEMENTAL PROBLEM

Sum Tangents

The sine and cosine functions are not the only ones for which there are angle sum formulas. The tangent function has this angle sum formula:

$$\tan(A+B) = \frac{\tan A + \tan B}{1 - \tan A \tan B}$$

Your task in this problem is to show how to prove this formula. One useful fact in proving this is the relationship $\tan\theta = \frac{\sin\theta}{\cos\theta}$.

1. **a.** Use the right-triangle definitions of the sine, cosine, and tangent functions to prove the relationship $\tan\theta = \frac{\sin\theta}{\cos\theta}$ for acute angles.

   b. Explain why this relationship holds for all angles.

2. Use the relationship $\tan\theta = \frac{\sin\theta}{\cos\theta}$ and the angle sum formulas for sine and cosine to prove the angle sum formula for the tangent function.

3. Develop an angle sum formula for the cotangent function similar to that for the tangent function. Here are two possible approaches.

   a. Apply the fact that $\cot\theta$ can be defined as $\frac{1}{\tan\theta}$ directly to the tangent-of-a-sum formula.

   b. Combine the fact that $\cot\theta$ is equal to $\frac{\cos\theta}{\sin\theta}$ with the angle sum formulas for sine and cosine.

Appendix A

SUPPLEMENTAL PROBLEM

Moving to the Second Quadrant

In *The Sine of a Sum,* you proved this sine-of-a-sum formula:

$$\sin(A + B) = \sin A \cos B + \cos A \sin B$$

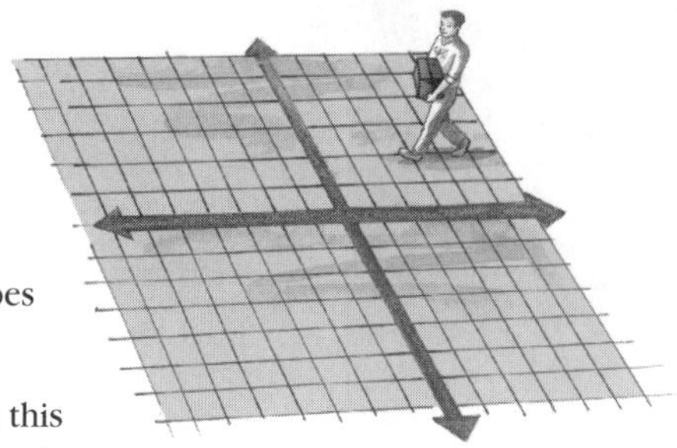

Unfortunately, the proof suggested in that activity depends on a diagram that requires A and B to be acute angles. So the proof in *The Sine of a Sum* does not apply in all cases.

Fortunately, the formula does hold for all angles. In this activity, you need to prove that it holds when A is in the second quadrant (and B is still in the first quadrant).

The basic idea is to express A as 90° more than some first-quadrant angle. To use this idea, you need to develop formulas connecting the sine and cosine of a second-quadrant angle with the sine and cosine of the first-quadrant angle that is 90° less.

1. The first step is to find a formula for $\sin(\theta + 90°)$ in terms of either $\sin\theta$ or $\cos\theta$.

   a. Pick a value for θ and find $\sin(\theta + 90°)$. Then find $\sin\theta$ and $\cos\theta$, and write a general formula based on what you find.

   b. Verify your formula from Question 1a using other values for θ.

Continued on next page

Interactive Mathematics Program   As the Cube Turns 199

c. Explain why your formula from Question 1a holds for all values of θ, in these ways:

- Using the graph from *Homework 14: Oh, Say What You Can See*
- Using the Ferris wheel model
- Using the general definition of the sine function in terms of coordinates

2. Develop and justify similar formulas for each of these expressions:

a. $\cos(\theta + 90°)$

b. $\sin(\theta - 90°)$

c. $\cos(\theta - 90°)$

3. Use your results from Questions 1 and 2 to find a formula for $\sin(A + B)$ for the case in which A is in the second quadrant but B is in the first quadrant.

Hints: Write A as $x + 90°$ and show that $\sin(A + B)$ is equal to $\sin[(x + B) + 90°]$. Then apply the formula from Question 1 to express $\sin[(x + B) + 90°]$ as a trigonometric function of $x + B$. Next, use the appropriate function-of-a-sum formula to express this function of $x + B$ in terms of the sine and cosine of x and B. Finally, get back to A by using the fact that x is $A - 90°$ and applying appropriate formulas from Question 2.

Appendix A

SUPPLEMENTAL PROBLEM

Adding 180°

In *Moving to the Second Quadrant*, you developed and explained formulas for the expressions sin (θ + 90°) and cos (θ + 90°) in terms of either sin θ or cos θ. This activity is similar, except that you will be adding 180° to θ instead of adding 90°.

1. Find a formula for sin (θ + 180°) in terms of either sin θ or cos θ, using these steps.

   a. Pick a value for θ and find sin (θ + 180°). Then find sin θ and cos θ, and write a general formula based on what you find.

   b. Verify your formula from Question 1a using other values for θ.

   c. Explain why your formula from Question 1a holds for all values of θ, in at least one of these ways:

   • Using the graph from *Homework 14: Oh, Say What You Can See*

   • Using the Ferris wheel model

   • Using the general definition of the sine function in terms of coordinates

2. Develop a similar formula for cos (θ + 180°) and justify your result.

3. Explain how to use your answers to Questions 1 and 2 to get formulas for sin (θ − 180°) and cos (θ − 180°). That is, show how to develop formulas in which you subtract 180° instead of adding 180°.

SUPPLEMENTAL PROBLEM

Sums for All Quadrants

In *The Sine of a Sum*, you showed that if A and B are first-quadrant angles, then $\sin(A + B)$ is equal to $\sin A \cos B + \cos A \sin B$. In *Moving to the Second Quadrant*, you showed that this formula works when one of the angles is in the second quadrant and the other is in the first quadrant.

Your task in this activity is to prove this formula for other cases.

1. Prove the formula for the case in which both A and B are second-quadrant angles.

   Hint: Adapt the method used in *Moving to the Second Quadrant* by writing A as $x + 90°$ and writing B as $y + 90°$. You will also need a formula for $\sin(\theta + 180°)$ in terms of either $\sin \theta$ or $\cos \theta$.

2. Prove the formula for all other cases, considering the possible combinations of quadrants for A and B. Use the periodicity of the sine and cosine functions, and express angles in various quadrants in terms of appropriate first-quadrant angles.

   Suggestion: Begin by listing the cases you need to consider, and think about how to avoid duplication. For instance, if you do the case in which A is in the third quadrant and B is in the second quadrant, you do not have to also do the case in which these quadrants are reversed.

Bugs in Trees

Matrices are basically a shorthand for representing certain numerical information, and matrix operations are a convenient way to describe certain arithmetic processes.

In this unit, matrices have been used as a way of doing the arithmetic involved in geometrical transformations, such as translations and rotations. In *Meadows or Malls?* (in Year 3), you used matrices to represent systems of linear equations.

This problem will illustrate another context in which matrices can provide a useful way to describe patterns of arithmetic operations.

The Situation

A pair of trees, tree A and tree B, are side by side. Some bugs have infested the trees, and the bugs are moving back and forth. A researcher watches the activities of the bugs, recording their movements once per hour, and reaches these conclusions:

- If a bug is in tree A at a particular observation, there is a 70% chance that it will still be in tree A at the next observation and only a 30% chance that it will be in tree B at the next observation.

- If a bug is in tree B at a particular observation, there is a 60% chance that it will be in tree A at the next observation and only a 40% chance that it will still be in tree B at the next observation.

Continued on next page

Comment: In studying sequences of coin flips, you learned that each flip is independent of the previous flips. In other words, the probability of getting heads or tails on a given flip does not depend on what was flipped in the past. In the situation here, the position of a bug at a given observation is analogous to the outcome of a given coin flip. But unlike the coin situation, a bug's position at a given observation is not independent of where it was before. Specifically (according to this researcher, at least), the position of a bug at a given observation depends (in a probabilistic way) on where it was at the preceding observation.

A situation in which each trial depends probabilistically on the preceding trial is called a *Markov chain*. The theory of Markov chains derives its name from the Russian probabilist A. A. Markov (1856-1922), who did pioneering work in this field.

Some Sample Questions

1. Suppose that in a certain observation, there are 110 bugs in tree A and 90 bugs in tree B. If the movement of the bugs follows the researcher's analysis, how many will be in each tree at the next observation?

2. Suppose the trees become infested with many thousands of bugs. At a given observation, 25% of the bugs are in tree A and 75% are in tree B. If these bugs continue to move according to the researcher's analysis, what fraction of them will be in each tree at the next observation?

Continued on next page

The Matrices

The researcher's probabilities can be put into this matrix:

$$\begin{bmatrix} .7 & .3 \\ .6 & .4 \end{bmatrix}$$

This is called a *transition matrix,* because it describes how the bug's position might change from one observation to the next.

The results of a particular observation can be put into a row matrix. For example, in Question 1, the information obtained can be represented by the row vector [110 90].

3. Show why the arithmetic you did in Question 1 can be represented in matrix form by the product

$$[110 \ 90] \begin{bmatrix} .7 & .3 \\ .6 & .4 \end{bmatrix}$$

4. Show how to represent the arithmetic of Question 2 as the product of a row vector and a matrix.

The Hours Go By

For the rest of this assignment, continue to assume that the researcher's analysis holds. Start with the situation from Question 2 and suppose that you make an observation every hour. Treat the initial situation as hour 0 and your result from Question 2 as hour 1.

5. Find the percentage of bugs in each tree at hour 2. (Remember that a bug's position at hour 2 depends on where it was at hour 1, but not on where it was at hour 0.)

Continued on next page

6. Develop a matrix expression for the distribution of the bugs at hour n.

7. What will happen in the long run? Will there ever come a time when the bugs are all in tree A? Explain your answers.

Adapted with permission from the *Mathematics Teacher*, © May 1998, by the National Council of Teachers of Mathematics.

SUPPLEMENTAL PROBLEM

Half a Sine

You have found formulas that allow you to find the sine and cosine of the sum and the difference of two angles in terms of the sines and cosines of the two angles themselves. You also found formulas for double angles in *Homework 18: Doubles and Differences*.

Your task in this problem is somewhat the opposite, namely, to find formulas for the sine and cosine of *half* of a given angle in terms of the sine and cosine of the angle itself.

In other words, you want to find formulas that look like this:

$$\sin \frac{A}{2} = \ldots$$
$$\cos \frac{A}{2} = \ldots$$

In each case, the right side of the equation can involve any trigonometric functions using the angle A itself.

Hint: Start by thinking of A as being twice $\frac{A}{2}$, and use the double-angle formula to write $\cos A$ in terms of $\sin \frac{A}{2}$ and $\cos \frac{A}{2}$. Then use the Pythagorean identity, $\sin^2 \theta + \cos^2 \theta = 1$, to get $\cos A$ in terms of only $\sin \frac{A}{2}$ or only $\cos \frac{A}{2}$, and work from that to get $\sin \frac{A}{2}$ and $\cos \frac{A}{2}$ in terms of $\cos A$.

SUPPLEMENTAL PROBLEM

The General Isometry

Mathematicians often use the word *transformation* to indicate a function in which the domain and range are sets of points. If we use the letter f to represent a specific transformation, then $f(X)$ represents the point to which X is moved.

An isometry is a special type of geometric transformation—one in which the distance between two points remains unchanged. In other words, if f is an isometry and A and B are any two points, then the distance from $f(A)$ to $f(B)$ must be equal to the distance from A to B.

You've also seen that there are three fundamental types of isometries of the plane: translations, rotations, and reflections. In this activity, you should suppose that ABC is a triangle and that f is an isometry of the plane. Your task in this activity is to show, as described in Questions 1 through 3, that f can be created by combining the three basic types of isometries.

1. Suppose RST is a triangle that is congruent to ABC, with $AB = RS$, $BC = ST$, and $AC = RT$. (*Reminder:* The notation XY means the distance from X to Y.)

Show that one of these two statements must be true.

- There is a translation g, a rotation h, and a reflection k for which $R = k(h(g(A)))$, $S = k(h(g(B)))$, and $T = k(h(g(C)))$.

Continued on next page

- There is a translation g and a rotation h for which $R = h(g(A))$, $S = h(g(B))$, and $T = h(g(C))$.

   Hint: First show that there is a translation g for which $g(A) = R$. Then show that there is a rotation h around R for which $h(g(B)) = S$. Finally, decide if a reflection through the line RS is needed.

2. Suppose RST is a triangle, and suppose that points X and Y satisfy these three conditions:

   - $RX = RY$
   - $SX = SY$
   - $TX = TY$

   Show that X and Y are actually the same point. (*Hint:* Show that if X and Y were different points, then R, S, and T would all be on the perpendicular bisector of $\overline{XY}$.)

3. Use your results from Questions 1 and 2 to prove that one of these two statements must be true.

   - There is a translation g, a rotation h, and a reflection k such that $f(X) = k(h(g(X)))$ for every point X.
   - There is a translation g and a rotation h such that $f(X) = h(g(X))$ for every point X.

   In other words, show that f is either a combination of a translation, a rotation, and a reflection or a combination of only a translation and a rotation.

SUPPLEMENTAL PROBLEM

Perspective on Geometry

One of the key tasks in this unit is deciding how to represent a three-dimensional object—the cube—on the two-dimensional calculator screen.

Artists use the general term *perspective* to describe the various methods they use to create this type of representation. Your task in this assignment is to do some research on the history of perspective in art.

Your report should describe different schemes of perspective and explain the geometric principles behind them. You may want to include examples of famous works of art showing different methods or create your own drawings illustrating how the same object might be drawn using different methods.

"Venice: A Regatta on the Grand Canal" by the Italian painter Canaletto (1697–1768).

SUPPLEMENTAL PROBLEM

Let the Calculator Do It!

In *Find Those Corners!* you projected the vertices of a cube on the plane $z = 5$. You probably found that computing the projections for each vertex was a lot of work.

In this assignment, you will write a program for your calculator (or a computer, if you and your teacher prefer) to do the work.

Your program should ask the user for the vertex of the cube or for any point that the user wants projected. It should also ask the user for the viewpoint. Your program should then tell the user the projection of the given point, using the given viewpoint, on the plane $z = 5$.

APPENDIX B

TURNCUBE

This program will run on the TI-82 calculator. When the program is run, the screen will show the cube turning (in 36 increments of 10° each). Here are the meanings of the key variables.

- Matrix *A* gives the vertices of the cube.
- Matrix *B* gives the coordinates of the projected vertices.
- Matrix *C* is the rotation matrix.
- The variables *J*, *K*, and *L* give the coordinates of the viewpoint.
- The variable *S* gives the location of the screen (as the plane $z = S$).
- The variables *R* and *Q* represent the ratio $\frac{z\text{-distance to the screen}}{z\text{-distance to the viewpoint}}$ for, respectively, the vertices on the back face of the cube (whose *z*-coordinate is 0) and the vertices on the front face of the cube (whose *z*-coordinate is 2).

The situation used in this program is specified in early lines of the program, as follows:

- The vertices of the cube are initially at $(1, 0, 0)$, $(1, 2, 0)$, $(3, 0, 0)$, $(3, 2, 0)$, $(1, 0, 2)$, $(1, 2, 2)$, $(3, 0, 2)$, and $(3, 2, 2)$.
- The viewpoint is at $(5, 4, 10)$.
- The screen is the plane $z = 4$.

This program is on separate pages so you can reproduce it by itself.

PROGRAM: TURNCUBE

: −3.7 → Xmin

: 5.7 → Xmax

: −2.1 → Ymin

: 4.1 → Ymax

: FnOff

: Degree

: [[1, 0, 0] [1, 2, 0] [3, 0, 0] [3, 2, 0] [1, 0, 2] [1, 2, 2] [3, 0, 2] [3, 2, 2]] → [A]

: [[cos 10, sin 10, 0] [−sin 10, cos 10, 0] [0, 0, 1]] → [C]

: 5 → J : 4 → K : 10 → L

: 4 → S

: S/L → R : (S − 2)/(L − 2) → Q

: {8, 2} → dim [B]

: For (C, 1, 37)

    : For (U, 1, 4)

        : [A] (U, 1) + R*(J − [A] (U, 1)) → [B] (U, 1)

        : [A] (U, 2) + R*(K − [A] (U, 2)) → [B] (U, 2)

    : End

    : For (U, 5, 8)

        : [A] (U, 1) + Q*(J − [A] (U, 1)) → [B] (U, 1)

        : [A] (U, 2) + Q*(K − [A] (U, 2)) → [B] (U, 2)

    : End

    : ClrDraw

    : Line ([B] (1, 1), [B] (1, 2), [B] (2, 1), [B] (2, 2))

    : Line ([B] (2, 1), [B] (2, 2), [B] (4, 1), [B] (4, 2))

    : Line ([B] (3, 1), [B] (3, 2), [B] (4, 1), [B] (4, 2))

    : Line ([B] (1, 1), [B] (1, 2), [B] (3, 1), [B] (3, 2))

    : Line ([B] (5, 1), [B] (5, 2), [B] (6, 1), [B] (6, 2))

    : Line ([B] (6, 1), [B] (6, 2), [B] (8, 1), [B] (8, 2))

    : Line ([B] (7, 1), [B] (7, 2), [B] (8, 1), [B] (8, 2))

    : Line ([B] (5, 1), [B] (5, 2), [B] (7, 1), [B] (7, 2))

: Line ([B] (1, 1), [B] (1, 2), [B] (5, 1), [B] (5, 2))
: Line ([B] (2, 1), [B] (2, 2), [B] (6, 1), [B] (6, 2))
: Line ([B] (3, 1), [B] (3, 2), [B] (7, 1), [B] (7, 2))
: Line ([B] (4, 1), [B] (4, 2), [B] (8, 1), [B] (8, 2))
: [A]*[C] → [A]
: End

Blackline Masters

This appendix contains these materials for the unit.

- The diagram showing a computer screen output, for use in developing the program using For/End loops on Day 3
- The graphs of the sine and cosine functions, for use in the discussion of *Homework 14: Oh, Say What You Can See*
- Two diagrams for use in the discussion of *Homework 20: "A Snack in the Middle" Revisited*
- The diagram of the three-dimensional coordinate system, for use on Day 23
- The in-class and take-home assessments for the unit

This appendix also contains a final assessment for the first semester of Year 4, suitable for schools using a traditional semester schedule. This assessment is designed on the assumption that your class will have completed *High Dive* and *As the Cube Turns*.

This semester assessment is not intended to be a comprehensive test of the material in these units, but focuses instead on some essential ideas. We recommend that you give students two hours to work on the semester assessment so they can complete it without time pressure, and that you allow them to use graphing calculators and to have access to their textbooks and notes (including portfolios).

Appendix C

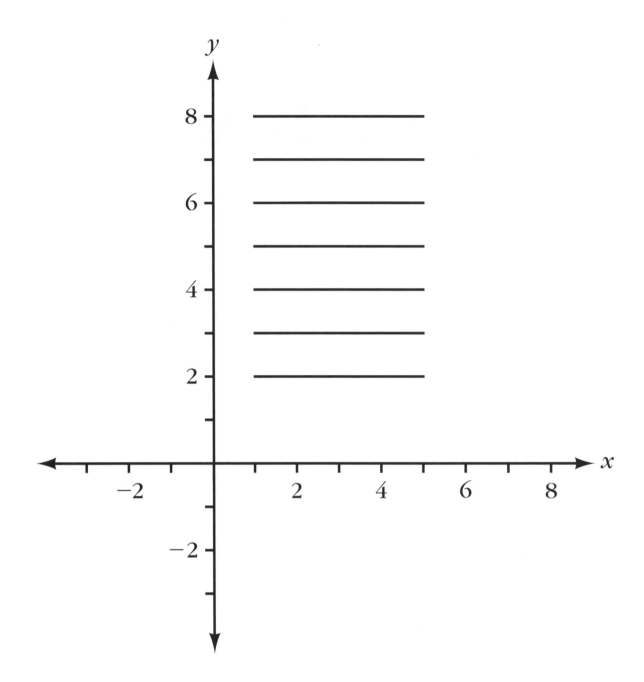

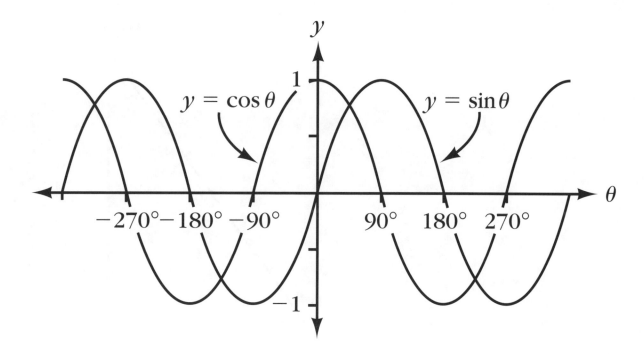

Appendix C

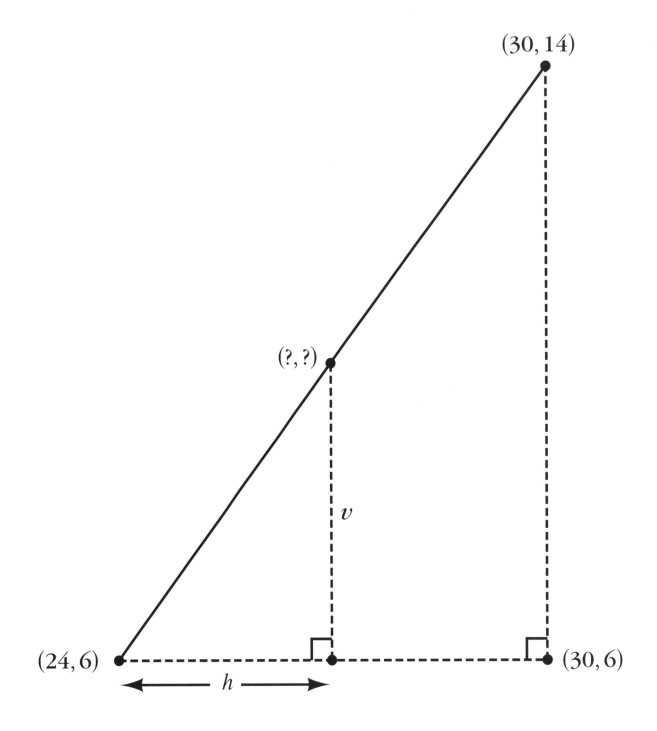

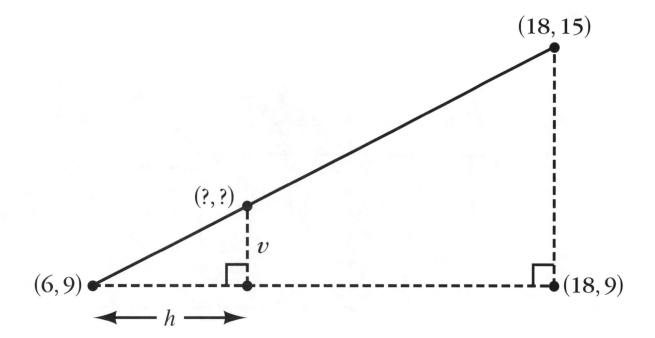

Appendix C

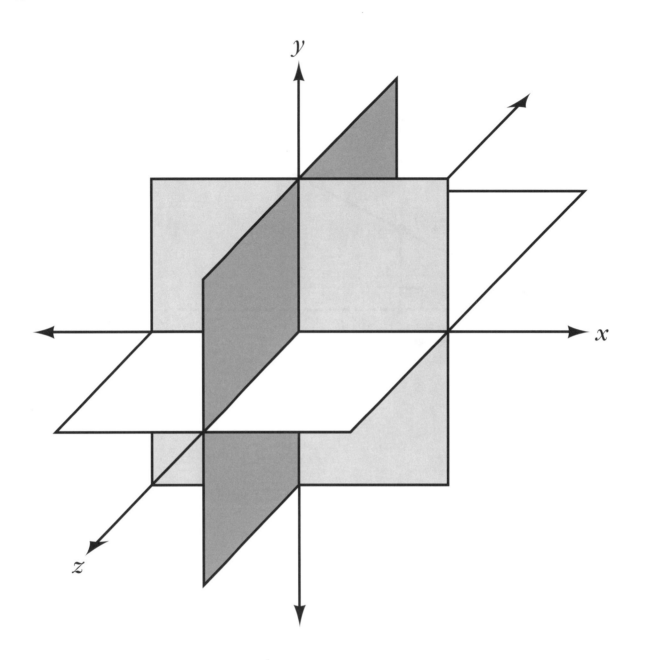

In-Class Assessment for *As the Cube Turns*

Here is a plain-language program for a graphing calculator:

Program: DRAW

Setup program

Let A be the matrix $\begin{bmatrix} 1 & 1 \\ 3 & 3 \\ 5 & 1 \end{bmatrix}$

Draw a line from (a_{11}, a_{12}) to (a_{21}, a_{22})

Draw a line from (a_{11}, a_{12}) to (a_{31}, a_{32})

Draw a line from (a_{31}, a_{32}) to (a_{21}, a_{22})

1. On a piece of graph paper, draw what the screen should look like after running this program using an appropriate viewing window. Show scales for the axes.

2. Use a loop and a matrix *B* to modify the program DRAW so the screen will look like this after running the new program.

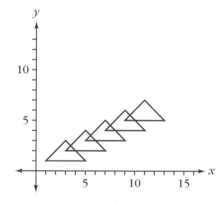

Continued on next page

3. Use a loop and a matrix *C* to modify the program DRAW so the screen will look like this after running the new program.

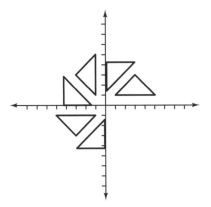

Note: The scales in this diagram are the same as in Question 2, although the viewing window is different. The numbers on the scales have been omitted to avoid cluttering the diagram.

Take-Home Assessment for *As the Cube Turns*

Part I: Where Are We?

This pair of equations will project a point (u, v, w) onto a screen parallel to the xy-plane so that the x- and y-coordinates of the projected point are a and b.

$$a = u + \left(\frac{6-w}{2-w}\right)(0 - u)$$

$$b = v + \left(\frac{6-w}{2-w}\right)(1 - v)$$

1. What are the coordinates of the viewpoint for the projection?

2. What is the equation of the screen?

3. Explain in detail why the equations given for a and b give the projection you describe.

Part II: How Does It Turn?

You have seen that when a point (x, y) is rotated counterclockwise around the origin through an angle ϕ, the new point has coordinates $x \cos \phi - y \sin \phi$ and $x \sin \phi + y \cos \phi$.

Explain how those two formulas were developed.

IMP Year 4 First Semester Assessment

I. Ferris Wheel Fence, Revisited

It's time to look back at the problem of the fence around the amusement park, from *High Dive*.

As you may recall, Al and Betty are riding on a Ferris wheel. This Ferris wheel has a radius of 30 feet, and its center is 35 feet above ground level.

There is a 25-foot-high fence around the amusement park, but once you get above the fence, there is a wonderful view.

What percentage of the time are Al and Betty above the level of the fence?

II. Opposite Angles

You have learned these formulas involving trigonometric functions:

$$\cos(-\theta) = \cos\theta$$

$$\sin(-\theta) = -\sin\theta$$

Explain each of these formulas in several ways:

- In terms of the Ferris wheel
- In terms of the graphs of the sine and cosine functions
- Using numerical examples

You can use these graphs of sine and cosine in your explanation:

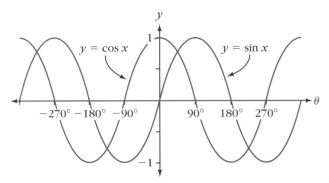

Continued on next page

You can also use this diagram to represent a Ferris wheel:

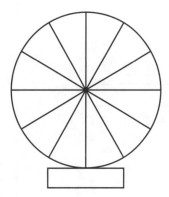

III. Clock Time

Imagine that the screen of your calculator is like the face of a clock with only a second hand. Think of the center of the clock and of the screen as the point (0, 0).

If the second hand were pointing to 12, the clock might look something like this:

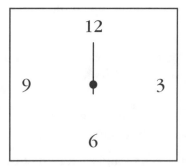

Your task is to write a program that shows the second hand turning, but your screen is going to have an *x*-axis and a *y*-axis on it, with (0, 0) at the center.

Your screen won't have the clock numbers or the dot in the center where the second hand is attached. Also, when the second hand is pointing straight up or down or straight to the right or left, you won't see the second hand because it will overlap with one of the axes. So, after the hand is pointing straight up, your next few screens should look something like the diagrams on the next page.

Continued on next page

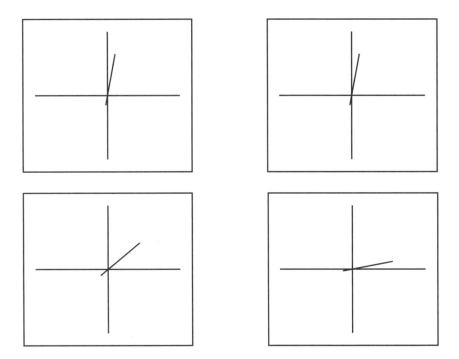

The sequence of diagrams should continue like this, as if the second hand is moving around the center of the screen.

Here are some details.

- Use a window setting in which the scales on the *x*- and *y*-axes are the same.
- Assume that when the second hand is pointing to 12, it is the line segment from $(0, -1)$ to $(0, 5)$.
- Have the second hand make two complete turns around the clock.

You should decide these things.

- How many degrees should each turn of the second hand be?
- How long a delay should there be between screens?

As an extra challenge, try to adjust the amount of each turn and the time of each delay so that it actually takes a minute for the hand to go all the way around.

GLOSSARY

Absolute value function family Informally, the family of functions whose graphs have the V-shape of the graph of the absolute value function defined by the equation $y = |x|$.

Example: The function defined by the equation $y = |2x + 1|$, whose graph is shown here, is considered a member of the absolute value function family.

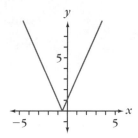

Acceleration See **velocity**.

Amplitude See **periodic function**.

Asymptote Informally, a line or curve to which the graph of an equation draws closer and closer as the independent variable approaches $+\infty$, approaches $-\infty$, or approaches a specific value.

Example: For the function f defined by the equation $f(x) = \frac{3}{x-1}$, both the horizontal line $y = 0$ (which is the x-axis) and the vertical line $x = 1$ (the dashed line in the diagram) are asymptotes. The graph approaches the line $y = 0$ as x approaches $+\infty$ or $-\infty$, and the graph approaches the line $x = 1$ as x approaches 1.

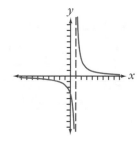

Interactive Mathematics Program

Bias   In sampling, the tendency of a sampling process to overrepresent or underrepresent a portion of the population being sampled. Avoiding bias is an important goal in sampling.

Binomial distribution   A probability distribution describing the result of repeated independent trials of the same event with two possible outcomes. If a particular outcome has probability p for each trial, the binomial distribution states that the probability that this outcome occurs exactly r times out of n trials is $_nC_r \cdot p^r \cdot (1-p)^{n-r}$, where $_nC_r$ is the combinatorial coefficient equal to $\frac{n!}{r!(n-r)!}$.

Example: Suppose a weighted coin has probability .7 of coming up heads. If the coin is flipped 50 times, the probability of getting exactly 30 heads is

$$_{50}C_{30} \cdot (.7)^{30} \cdot (.3)^{20}$$

Central angle   An angle formed at the center of a circle by two radii.

Example: The angle labeled θ is a central angle.

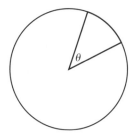

Central limit theorem   A statistical principle stating that when samples of a sufficiently large given size are taken from almost any population, the means of these samples are approximately normally distributed. See *The Central Limit Theorem* in *The Pollster's Dilemma*.

Circular function   Any of several functions defined by placing an angle θ with its vertex at the origin and the initial ray of the angle along the positive x-axis, as shown in the diagram on the next page. If (x, y) is any point different from $(0, 0)$ on the terminal ray of the angle, then the sine, cosine, and tangent of θ are defined by the equations

$$\sin \theta = \frac{y}{r}$$

$$\cos \theta = \frac{x}{r}$$

$$\tan \theta = \frac{y}{x}$$

where $r = \sqrt{x^2 + y^2}$

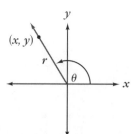

The circular functions also include the secant, cosecant, and cotangent functions, which are defined as other ratios involving *x, y,* and *r.* (The circular functions are also called *trigonometric functions.*)

Commutative   An operation ∗ is commutative if the equation $a * b = b * a$ holds true for all values of *a* and *b*. An operation that does not have this property is called *noncommutative*.

Example: The operation of addition is commutative because for any numbers *a* and *b*, $a + b = b + a$. The operation of subtraction is noncommutative because, for example, $5 - 9 \neq 9 - 5$.

Completing the square   The process of adding a constant term to a quadratic expression so that the resulting expression is a perfect square.

Example: To complete the square for the expression $x^2 + 6x + 4$, add 5 to get $x^2 + 6x + 9$, which is the perfect square $(x + 3)^2$.

Complex number   A number of the form $a + bi$, where *a* and *b* are real numbers and *i* is the number $\sqrt{-1}$. The number *a* is the *real part* of $a + bi$, and *bi* is its *imaginary part*.

Components of velocity   The velocity of an object is sometimes expressed in terms of separate *components of velocity*. For instance, a falling object that is also moving to the side has both a vertical and a horizontal component of velocity.

Example: In the diagram on the next page, if a person is swimming across a river at a speed of 5 feet per second, in the direction shown by the solid line, then the "toward-shore" component of the swimmer's velocity is 5 sin 60° feet per second and the "parallel-to-shore" component of the swimmer's velocity is 5 cos 60° feet per second. (The "toward-shore" component can also be expressed as 5 cos 30° feet per second.)

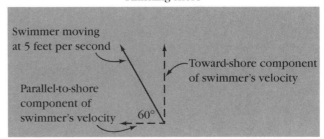

Composition of functions

An operation on functions in which the output from one function is used as the input for another. The composition of two functions f and g is written $f \circ g$, and this function is defined by the equation $(f \circ g)(x) = f(g(x))$. A function formed in this manner is called a *composite function*.

Example: Suppose f and g are defined by the equations $f(x) = x^2$ and $g(x) = 2x + 1$. Then $f \circ g$ is defined by the equation $(f \circ g)(x) = (2x + 1)^2$.

Confidence interval

In a sampling process from a population, an interval around the sample mean that has a certain likelihood, called the *confidence level*, of containing the true mean for the population. If the interval is symmetric around the sample mean, then half the length of the interval is called the **margin of error.**

Example: The interval of values within one standard deviation of the sample mean is a *95% confidence interval*, because if that procedure is followed for many samples, it will contain the true mean 95% of the time.

Conic section

Any of several two-dimensional figures that can be formed by the intersection of a plane with a cone. The specific figure formed generally depends on the angle at which the plane meets the cone, and is "usually" an ellipse, parabola, or hyperbola.

Constant function — A function whose value is the same for every input.

Correlation coefficient — A number between −1 and 1, usually labeled r, that measures how well a set of data pairs can be fitted by a linear function. The closer r is to 1, the better the data set can be fitted by a linear function with positive slope; the closer r is to −1, the better the data set can be fitted by a linear equation with negative slope. If r is not close to either 1 or −1, then the data set cannot be approximated well by any linear function.

Cosine — See **circular function.**

Cubic function — A function defined by an equation of the form $y = ax^3 + bx^2 + cx + d$ in which a, b, c, and d are real numbers with $a \neq 0$.

Dependent variable — See **function.**

Directly proportional — A relationship between two quantities or variables in which one of the variables is a constant multiple of the other variable.

Example: If an object has been traveling at 20 miles per hour, then the distance it has traveled is directly proportional to the amount of time it has been traveling. This can be seen algebraically as follows: If d represents the distance (in miles) and t represents the time elapsed (in hours), then d and t satisfy the equation $d = 20t$.

Discriminant — See **quadratic formula.**

Domain — The set of values that can be used as inputs for a given function.

Example: If f is the function defined by the equation $f(x) = \frac{x}{x^2 - 1}$, then the domain of f is the set of all real numbers except 1 and −1.

Year 4

End behavior   The behavior of a function as the independent variable approaches $+\infty$ or $-\infty$. See *Approaching Infinity* in *The World of Functions*.

Example: For the function defined by the equation $y = 2^x$, the end behavior is that y increases without bound as x approaches $+\infty$ and that y approaches 0 as x approaches $-\infty$.

Exponential function family   The family of functions defined by equations of the form $y = a \cdot b^x$ in which a is a nonzero real number and b is a positive real number other than 1. If $b > 1$, the function is an *exponential growth* function; if $b < 1$, the function is an *exponential decay* function. Functions in this family are characterized by the property that a fixed change in the independent variable always results in the same *percentage* change in the dependent variable.

Fibonacci sequence   The numerical sequence $1, 1, 2, 3, 5, 8, 13, \ldots$, in which the first two terms are both 1 and each succeeding term is the sum of the two preceding terms. (For example, the sixth term, 8, is the sum of the fourth and fifth terms, 3 and 5.) The sequence is often represented using the *recursion equation*
$$a_{n+2} = a_{n+1} + a_n$$

Function   Informally, a relationship in which the value of one variable (the **independent variable**) determines the value of another (the **dependent variable**). In terms of an In-Out table, the independent variable gives the input and the dependent variable gives the output. In terms of a graph, the independent variable is generally shown on the horizontal axis and the dependent variable on the vertical axis.

Formally, a function is a set of number pairs for which two different pairs cannot have the same first coordinate.

Example: The equation $y = x^2$ expresses y as a function of x.

Greatest integer function	See **step function**.
Identity	1. An equation that holds true no matter what numbers are substituted for the variables (as long as the expressions on both sides of the equation make sense).

Example: The equation $(a + b)^2 = a^2 + 2ab + b^2$ is an identity, because this equation holds true for all real numbers a and b.

2. For a given operation, an *identity* (or an *identity element*) for that operation is an element which, when combined with any element using the given operation, yields that second element as the result.

Examples: The number 0 is the identity for addition because $x + 0$ and $0 + x$ are both equal to x for any number x. Similarly, the number 1 is the identity for multiplication.

Identity function	The function on a given domain whose output is equal to its input. This function is the identity for the operation of composition of functions and can be represented by the equation $f(x) = x$.
Independent events	Two (or more) events are independent if the outcome of one does not affect the outcome of the other.
Independent variable	See **function**.
Inverse	If an operation has an **identity,** then an *inverse* for a given element (under that operation) is an element which, when combined with the given one, yields the identity as the result.

Examples: The number -7 is the inverse for 7 for the operation of addition because both $7 + (-7)$ and $(-7) + 7$ are equal to 0, which is the identity for addition. Similarly, the number $\frac{1}{5}$ is the inverse for 5 for the operation of multiplication.

Inverse of a function

Informally, the inverse of a function f is the function that "undoes" f. Formally, the inverse of f is its inverse with regard to the operation of composition. That is, the inverse of f is the function g for which both $f \circ g$ and $g \circ f$ are the appropriate identity functions, with $(f \circ g)(x) = x$ and $(g \circ f)(x) = x$. The inverse of f is sometimes represented by f^{-1}.

Example: If f is the function defined by the equation $f(x) = 3x + 2$, then the inverse of f is the function g given by the equation $g(x) = \frac{x-2}{3}$. The fact that g "undoes" f is illustrated by the fact that $f(5) = 17$ and $g(17) = 5$. In terms of composition, we have $(f \circ g)(17) = 17$ and $(g \circ f)(5) = 5$.

Inversely proportional

A relationship between two quantities or variables in which one of the variables is obtained by dividing some constant by the other variable.

Example: If the length and width of a rectangle are to be chosen so that the rectangle has an area of 30 square inches, then the length will be inversely proportional to the width. This can be shown algebraically as follows: If L represents the length and W represents the width (both in inches), then L and W satisfy the equation $LW = 30$, so $L = \frac{30}{W}$.

Isometry

A geometrical transformation T with the property that for any pair of points A and B, the distance between $T(A)$ and $T(B)$ is equal to the distance between A and B. Isometries do not change the size or shape of geometric figures, and include **translations, rotations,** and **reflections.**

Least squares method

A method of determining a function from a given family that best fits a set of data. For a finite set of data points (x_1, y_1), $(x_2, y_2), \ldots, (x_n, y_n)$, the least-squares method seeks the function f (from the family) that minimizes the value of the expression

$$\sum_{i=1}^{n} [y_i - f(x_i)]^2$$

in which $y_i - f(x_i)$ represents the vertical distance between the graph of f and the data point (x_i, y_i).

Linear equation	For one variable, an equation of the form $ax + b = 0$, in which a and b are real numbers with $a \neq 0$. For n variables $x_1, x_2, \ldots, x_n$, an equation of the form $a_1 x_1 + a_2 x_2 + \ldots + a_n x_n + b = 0$.
Linear function	For a function with one input variable, a function defined by an equation of the form $y = ax + b$ in which a and b are real numbers. (The special case of a constant function, in which $a = 0$, is sometimes excluded from the family of linear functions.) The definition is similar for functions with more than one input variable.
Linear regression	A process for obtaining the linear function that best fits a set of data. The equation for this function is the *regression equation* and its graph is the *regression line*.
Logarithmic function family	Informally, the family of functions defined by equations of the form $y = a + b \log x$ (or of the form $y = a + b \ln x$), where a and b are real numbers with $b \neq 0$.
Loop	In programming, a set of instructions that specifies the repeated execution of a given set of steps.
Margin of error	See **confidence interval.**
Mean of a discrete probability distribution	If a probability distribution has possible outcomes $x_1, x_2, \ldots, x_n$, and the outcome x_i has probability $P(x_i)$, then the mean of the distribution is given by the expression $$\sum_{i=1}^{n} P(x_i) \cdot x_i$$ The mean of the distribution is numerically equal to the expected value of the event that the distribution describes. See *Mean and Standard Deviation for Probability Distributions* in *The Pollster's Dilemma*.
Nested loop	A programming loop that occurs within the body of another loop.

Year 4

Normal curve   The graph that represents a normal distribution. If the normal distribution has mean μ and standard deviation σ, then the equation of its graph is

$$y = \left(\frac{1}{\sigma\sqrt{2\pi}}\right) \cdot e^{-\frac{1}{2}\left(\frac{x-\mu}{\sigma}\right)^2}$$

Example: If $\mu = 0$ and $\sigma = 1$, the equation simplifies to

$$y = \left(\frac{1}{\sqrt{2\pi}}\right) \cdot e^{-\frac{1}{2}x^2}$$

This special case is called the *standard normal curve*. The diagram here shows the graph of this equation.

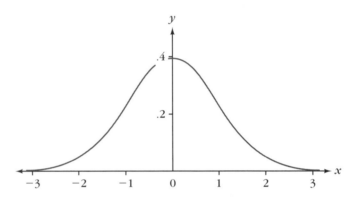

Parabola   The general shape for the graph of a quadratic function. Also see **conic section.**

Parameter   Often, a variable whose value is used to specify a particular member of a family of functions. (The term *parameter* has other meanings as well.)

Example: The family of quadratic functions consists of all functions defined by equations of the form $y = ax^2 + bx + c$, where a, b, and c are real numbers with $a \neq 0$. The variables a, b, and c are parameters whose numerical values specify a particular quadratic function. The set of quadratic functions is called a *three-parameter family*.

Period   See **periodic function.**

Periodic function   Informally, a function whose values repeat after a specific interval. Specifically, a function f is periodic if there is a positive number a such that $f(x + a) = f(x)$ for all values of x. The smallest positive value for a is called the **period** of f.

If a periodic function has a maximum and a minimum value, then half the difference between these values is the **amplitude** of the function.

Example: The function f defined by the equation $f(x) = 3 \sin (2x) + 5$ is a periodic function with period 180°, because $f(x + 180°) = f(x)$ for all values of x. The graph shown here for this function illustrates its periodic behavior. The maximum value for f is 8 [for instance, $f(45°) = 8$] and the minimum value is 2 [for instance, $f(135°) = 2$], so the amplitude of f is $\frac{1}{2}(8 - 2) = 3$.

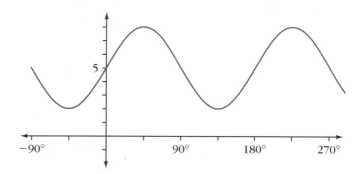

Polar coordinates   A system in which a point in the plane is identified by means of a pair of coordinates (r, θ) where r is the distance from the origin to the point and θ is the angle between the positive x-axis and the ray from the origin through the point, measured counterclockwise. See *A Polar Summary* in *High Dive*.

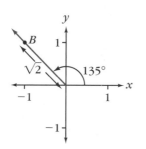

Example: The point B in the diagram, whose rectangular coordinates are $(-1, 1)$, has polar coordinates $(\sqrt{2}, 135°)$ because the distance from $(0, 0)$ to B is $\sqrt{2}$ and the angle from the positive x-axis to the ray through B is 135°. [*Note:* The point B has other polar coordinate representations, such as $(-\sqrt{2}, 315°)$ or $(\sqrt{2}, 495°)$.]

Year 4

Polynomial function

A function defined by an equation of the form $y = a_n x^n + a_{n-1} x^{n-1} + \ldots + a_1 x + a_0$, in which $a_n, a_{n-1}, \ldots, a_1,$ and a_0 are real numbers. If $a_n \neq 0$, the polynomial has *degree n*. The family of polynomial functions includes **constant functions** (degree 0), **linear functions** (degree 1), **quadratic functions** (degree 2), and **cubic functions** (degree 3), as well as functions of higher degree.

Example: The function defined by the equation $y = 3x^4 - 2x^2 + 5x - 1$ is a polynomial function of degree 4.

Power function

Generally, a function defined by an equation of the form $y = ax^b$ in which a and b are real numbers. (In some contexts, restrictions are imposed on b, such as requiring that b be an integer.)

Example: The function defined by the equation $y = 5x^{\frac{1}{2}}$ is a power function.

Principal value

A term sometimes used in the definitions of the inverse trigonometric functions to identify a specific number whose sine, cosine, or tangent is a given value.

Example: The equation $\sin x = 0.5$ has infinitely many solutions, but the solution $x = 30°$ is selected as the principal value, so that $\sin^{-1}(0.5)$ is defined to be $30°$.

Probability distribution

A set of values giving the probability for each possible outcome for an event.

Example: If a fair coin is flipped twice and we count the number of heads, the probability distribution is $P(2 \text{ heads}) = \frac{1}{4}, P(1 \text{ head}) = \frac{1}{2}, P(0 \text{ heads}) = \frac{1}{4}$.

Projection

A process for representing a three-dimensional object by means of a two-dimensional figure, or any representation of a figure by a lower-dimensional figure.

Pythagorean identity
Any of several trigonometric identities based on the Pythagorean theorem.

Example: The equation $\sin^2 x + \cos^2 x = 1$, which holds true for all values of x, is a Pythagorean identity.

Quadratic equation
For one variable, an equation of the form $ax^2 + bx + c = 0$ in which a, b, and c are real numbers with $a \neq 0$.

Quadratic formula
A formula for finding the solutions to a quadratic equation in terms of the coefficients. Specifically, for the quadratic equation $ax^2 + bx + c = 0$ (where a, b, and c are real numbers with $a \neq 0$), the solutions are given by the quadratic formula expression $\frac{-b \pm \sqrt{b^2 - 4ac}}{2a}$. The expression $b^2 - 4ac$ in this formula is called the **discriminant** of the equation.

Quadratic function
A function defined by an equation of the form $y = ax^2 + bx + c$ in which a, b, and c are real numbers with $a \neq 0$.

Radian
The measure of a central angle of a circle that intercepts a portion of the circumference whose length is equal to the radius of the circle. A radian is approximately equal to 57°.

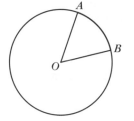

Example: In this diagram, the length of the arc from A to B is equal to the length of the radius $\overline{OA}$ of the circle, so angle AOB measures one radian.

Random
A term used in probability to indicate that any of several events is equally likely or, more generally, that an event is selected from a set of events according to a precisely described probability distribution.

Range
The set of values that can occur as outputs for a given function.

Example: If f is the function defined by the equation $f(x) = x^2$, then the range of f is the set of all nonnegative real numbers.

Rational function A function that can be expressed as the quotient of two polynomial functions.

Examples: The function defined by the equation $y = \frac{3}{x}$ is a rational function, as is the function defined by the equation $y = \frac{x^2 + 3x - 7}{2x^3 + 4x + 1}$. A polynomial function is a special type of rational function. For instance, the function defined by the equation $y = x^3 - 2x^2 + 3$ can be expressed as a quotient of polynomials by writing the equation as $y = \frac{x^3 - 2x^2 + 3}{1}$.

Rectangular coordinates A system in which a point is identified by coordinates that give its position in relation to each of the mutually perpendicular coordinate axes. In the plane, these axes are usually called the *x*-axis (horizontal) and *y*-axis (vertical). The horizontal coordinate is given first.

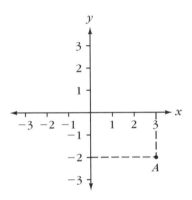

Example: In this diagram, point A has coordinates $(3, -2)$ because it corresponds to the number 3 on the *x*-axis and the number -2 on the *y*-axis.

Recursion A process for defining a sequence of numbers by specifying how to obtain each term from the preceding term(s). An equation specifying how each term is defined in this way is called a *recursion equation*.

Example: The sequence $1, 3, 5, 7, \ldots$ (the sequence of positive odd integers) can be defined by the recursion equation $a_{n+1} = a_n + 2$ together with the *initial condition* that $a_1 = 1$.

Reflection   A type of isometry in which the output for each point is its mirror image. A reflection is sometimes called a "flip."

Example: The diagram here illustrates a reflection in which the *y*-axis is a *line of reflection*. The triangle in the second quadrant is the reflection of the triangle in the first quadrant. In three dimensions, the role of a line of reflection is replaced by a *plane of reflection*.

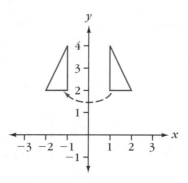

Rotation   A type of isometry in which the output for each point is obtained by rotating the point through an angle of a given size around a given point.

Example: The diagram below illustrates a 90° rotation counterclockwise around the origin. Each point of the lightly shaded triangle is moved to a point that is the same distance from the origin as the original point, but which is 90° counterclockwise (with respect to the origin) from the original point. The diagram shows the "paths" of two of the vertices of the lightly shaded triangle.

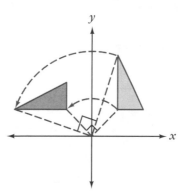

Sample mean   The mean for a sample taken from a population. The sample mean is often used to estimate the **true mean** for the population.

Year 4

Sample proportion

In a sample from a population, the fraction of sampled items that represent a specific outcome. The sample proportion is often used to estimate the **true proportion.**

Example: If an election poll of 500 voters shows 285 voters favoring a certain candidate, the sample proportion for the poll is $\frac{285}{500}$, which may be expressed as .57 or as 57%.

Sampling

A process of selecting members of a population and studying their characteristics in order to estimate or predict certain characteristics of the entire population. The selected members of the population comprise the *sample*.

In selecting the sample at random from a population, we sometimes consider a member of the population ineligible for further selection once it has already been selected. This is called *sampling without replacement*. If members of the population are eligible for repeated selection, this is called *sampling with replacement*.

Sine

See **circular function.**

Sine family of functions

The family of functions whose graphs have the same shape as the graph of the sine function. These functions can be written in the form $y = a \sin(bx + c) + d$, where a, b, c, and d are real numbers with a and b not equal to 0. This family includes the cosine function, because $\cos x = \sin\left(x + \frac{\pi}{2}\right)$.

Standard deviation of a discrete probability distribution

If a probability distribution has possible outcomes $x_1, x_2, \ldots, x_n$, and the outcome x_i has probability $P(x_i)$, then the standard deviation of the distribution is given by the expression

$$\sqrt{\sum_{i=1}^{n} P(x_i) \cdot (x_i - \mu)^2}$$

where μ is the mean of the distribution. See *Mean and Standard Deviation for Probability Distributions* in *The Pollster's Dilemma*.

Step function   Informally, a function whose graph consists of horizontal line segments.

Example: The **greatest integer function,** written $[x]$, is defined by the condition that $[x]$ is the largest integer N such that $N \leq x$. The diagram here shows the graph of this step function.

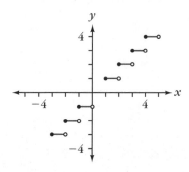

Tangent   See **circular function.**

Transformation
1. A *transformation of a function* is any of certain changes to a function that shift the graph vertically or horizontally or that stretch or shrink the graph vertically or horizontally.

2. A *geometrical transformation* is a function whose domain is the set of the points in the plane or the set of points in 3-space, in which the image of each point is another point (possibly the same point). An **isometry** is a special type of geometrical transformation.

Translation   A type of isometry in which the output for each point is obtained by moving the point a fixed amount in each of the coordinate directions. A translation is sometimes called a "slide."

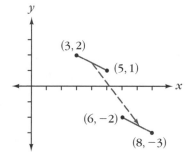

Example: In the diagram here, the line segment connecting $(3, 2)$ and $(5, 1)$ is being translated 3 units to the right and 4 units down, so that its image under the translation is the line segment connecting $(6, -2)$ and $(8, -3)$.

True mean — The mean for a population. See also **sample mean.**

True proportion — In a population, the fraction of items that represent a specific outcome. See also **sample proportion.**

Example: If 600 out of 1000 balls in an urn are red, then the true proportion of red balls is $\frac{600}{1000}$, which may be expressed as .6 or as 60%.

Unit circle — A circle of radius 1, especially the circle in the coordinate plane with radius 1 and center at the origin.

Variance — The square of standard deviation. For a finite set of data $x_1, x_2, x_3, \ldots, x_n$ with mean μ, the variance is given by the expression

$$\frac{1}{n} \sum_{i=1}^{n} (x_i - \mu)^2$$

Velocity — A combination of the speed and direction of a moving object. If an object is moving vertically, we generally consider upward motion as having a positive velocity and downward motion as having a negative velocity. The rate at which the velocity of an object is changing is the object's **acceleration.** A positive acceleration corresponds to an increase in velocity.

PHOTOGRAPHIC CREDITS

Teacher Book Interior Photography

ix Lynne Alper; **xi** FPG International; **xii** Black Star; **9** Eaglecrest High School, CO; Michelle Novotny; **30** Capuchino High School, CA; Peter Jonnard, Hillary Turner, Richard Wheeler; **43** Capuchino High School, CA; Peter Jonnard, Hillary Turner, Richard Wheeler; **65** Lynne Alper; **73** Lynne Alper; **95** Capuchino High School, CA; Peter Jonnard, Hillary Turner, Richard Wheeler; **109** Capuchino High School, CA; Dean Orfanedes, Hillary Turner, Richard Wheeler; **214** Lynne Alper; **243** Brookline High School, MA; Terry Nowak, Lynne Alper.

Student Book Interior Photography

3 Oxnard High School, CA; Jerry Neidenbach; **5** Tony Stone Images; **7** Hillary Turner; **14** Capuchino High School, CA; Chicha Lynch, Hillary Turner, Richard Wheeler; **20** Tony Stone Images; **23** PhotoEdit; **27** PhotoDisc; **30** Capuchino High School, CA; Chicha Lynch, Hillary Turner, Richard Wheeler; **34** SuperStock, Inc.; **37** FPG International; **38** Capuchino High School, CA; Peter Jonnard, Hillary Turner, Richard Wheeler; **41** Tony Stone Images; **43** The Image Bank; **44** FPG International; **45** FPG International; **48** Capuchino High School, CA; Peter Jonnard, Hillary Turner, Richard Wheeler; **52** PhotoDisc; **53** SuperStock, Inc.; **54** Capuchino High School, CA; Chicha Lynch, Hillary Turner, Richard Wheeler; **55** Hillary Turner; **62** Stock Boston; **68** Capuchino High School, CA; Chicha Lynch, Hillary Turner, Richard Wheeler; **70** Tony Stone Images; **72** Tony Stone Images; **78** Capuchino High School, CA; Peter Jonnard, Hillary Turner, Richard Wheeler; **77** The Image Bank; **79** The Image Bank; **80** PhotoEdit; **82** SuperStock, Inc.; **85** Leo de Wys, Inc.; Tony Stone Images; **87** FPG International; **91** Capuchino High School, CA; Peter Jonnard, Hillary Turner, Richard Wheeler; **93** SuperStock, Inc.; **98** Corbis/Bettmann; **99** FPG International; **102** Leo de Wys, Inc.; **104** SuperStock, Inc.; **110** Leo de Wys, Inc.; **113** Capuchino High School, CA; Peter Jonnard, Hillary Turner, Richard Wheeler; **114** Hillary Turner; **118** Hillary Turner; **119** Hillary Turner; **121** Aptos High School, CA; Anthony Pepperdine; **124** Hillary Turner; **129** The Image Bank; **130** SuperStock, Inc; **131** SuperStock, Inc; **133** Brookline High School, MA; Terry Nowak, Lynne Alper; **140** SuperStock, Inc.; **142** Capuchino High School, CA; Peter Jonnard, Hillary Turner, Richard Wheeler; **143** Stock Boston; **149** Tony Stone Images; **150** Stock Boston; **153** Stock Boston; **159** Shasta High School, CA; Dave Robathan; **161** Leo de Wys, Inc.; **163** FPG International; **176** SuperStock, Inc.; **182** PhotoDisc; **184** Capuchino High School, CA; Chicha Lynch, Hillary Turner and Richard Wheeler; **187** SuperStock, Inc.; **188** The Image Bank; **190** Foothill High School, CA; Cheryl Dozier; **208** Animals, Animals; **210** The Image Works; **215** Capuchino High School, CA; Dean Orfanedes; **216** FPG International; **222** Hillary Turner; **233** Hillary Turner, Richard Wheeler; **236** Corbis/Bettmann; **241** Hillary Turner, Richard Wheeler; **242** Brookline High School, MA; Terry Nowak, Lynne Alper; **243** Hillary Turner; **257** Foothill High School, CA; Cheryl Dozier; **258** Tony Stone Images; **259** Leo de Wys, Inc.; **262** Palm Press/©Harold E. Edgerton; **265** Capuchino High School, CA; Dean Orfanedes, Hillary Turner, Richard Wheeler; **280** FPG International; **281** FPG International; **282** Capuchino High School, CA; Dean Orfanedes,

Hillary Turner, Richard Wheeler; **284** Hillary Turner; **291** Capuchino High School, CA; Chicha Lynch; **292** Corbis; **293** PhotoEdit; **300** San Lorenzo Valley High School, CA; Sandie Gilliam, Lynne Alper; **307** PhotoEdit; **308** Capuchino High School, CA; Peter Jonnard, Hillary Turner, Richard Wheeler; **310** Hillary Turner; **312** Tony Stone Images; **314** Hillary Turner, Richard Wheeler; **315** San Lorenzo Valley High School, CA; Sandie Gilliam, Lynne Alper; **320** SuperStock, Inc.; **335** San Lorenzo Valley High School, CA; Sandie Gilliam, Lynne Alper; **341** Foothill High School, CA; Madeline Rippe, Cheryl Dozier; **348** Leo de Wys, Inc.; **357** The Image Bank; **359** Comstock; **361** PhotoDisc; **365** Capuchino High School, CA; Chicha Lynch, Hillary Turner, Richard Wheeler; **373** FPG International; **377** Brookline High School, MA; Terry Nowak; **386** Capuchino High School, CA; Chicha Lynch, Hillary Turner, Richard Wheeler; **395** Hillary Turner; **400** PhotoEdit; **409** Santa Cruz High School, CA; George Martinez, Lynne Alper; **414** SuperStock, Inc.; **415** SuperStock, Inc.; **416** FPG International; **417** The Image Bank; **420** SuperStock, Inc.; **421** SuperStock, Inc.; **423** Patrick Henry High School, MN; Jane Kostik; **429** The Image Bank; **430** Black Star; **431** SuperStock, Inc.; **434** The Image Works; **436** Foothill High School, CA; Cheryl Dozier; **445** FPG International; **447** The Image Bank.

Cover Photography

High Dive Corbis and Comstock; *Know How* Tony Stone Images, Inc.; *As the Cube Turns* Hillary Turner and Rick Helf; *The World of Functions* Hillary Turner and Richard Wheeler; *The Pollster's Dilemma* Corbis; *Back cover* Shasta High School, CA; Dave Robathan.

Front Cover Students

The World of Functions first row: Hilda Chavez, Mary Truong, Tom Hitchner, Enrique Gonzales. Second Row: Kermit Bayless, Jr., Kei Takeda, Ryan Alexander-Tanner, Rena Davis.

Interactive Mathematics Program
Year 4
Comment Form

Please take a moment to provide us with feedback about IMP. If you have comments or suggestions about Year 4, we'd like to read them. Once you've filled out this form, all you have to do is fold it and drop it in the mail. We'll pay the postage. Thank you!

Your Name _____

School _____

School Address _____

City/State/Zip _____

Phone _____

1. List any comments about the IMP *Year 4* student text.

   _____
   _____
   _____
   _____
   _____
   _____

2. List any comments about the teacher's guide for _____.

   _____
   _____
   _____
   _____

3. Do you have any other comments about IMP *Year 4* or any suggestions for improving the student text or teacher material?

   _____
   _____
   _____
   _____
   _____

Thank you for taking the time to fill out this comment form.
Please return completed forms to:
   Editorial—IMP, Key Curriculum Press, Box 2304, Berkeley, CA 94702.
   You can fold this form as shown on the back and it becomes a postage-paid self-mailer!

Fold carefully along this line.

BUSINESS REPLY MAIL
FIRST CLASS MAIL   PERMIT NO. 151   BERKELEY, CA

POSTAGE WILL BE PAID BY ADDRESSEE

KEY CURRICULUM PRESS
Innovators in Mathematics Education

P.O. Box 2304
Berkeley, CA 94702-9983
Attn: Editorial—IMP

NO POSTAGE
NECESSARY
IF MAILED
IN THE
UNITED STATES

Fold carefully along this line.

Interactive Mathematics Program
Year 4
Comment Form

Please take a moment to provide us with feedback about IMP. If you have comments or suggestions about Year 4, we'd like to read them. Once you've filled out this form, all you have to do is fold it and drop it in the mail. We'll pay the postage. Thank you!

Your Name _____

School _____

School Address _____

City/State/Zip _____

Phone _____

1. List any comments about the IMP *Year 4* student text.

   _____
   _____
   _____
   _____
   _____
   _____

2. List any comments about the teacher's guide for _____.

   _____
   _____
   _____
   _____
   _____

3. Do you have any other comments about IMP *Year 4* or any suggestions for improving the student text or teacher material?

   _____
   _____
   _____
   _____
   _____
   _____

Thank you for taking the time to fill out this comment form.
Please return completed forms to:
   Editorial—IMP, Key Curriculum Press, Box 2304, Berkeley, CA 94702.
   You can fold this form as shown on the back and it becomes a postage-paid self-mailer!

Fold carefully along this line.

BUSINESS REPLY MAIL		
FIRST CLASS MAIL	PERMIT NO. 151	BERKELEY, CA

POSTAGE WILL BE PAID BY ADDRESSEE

P.O. Box 2304
Berkeley, CA 94702-9983
Attn: Editorial—IMP

NO POSTAGE
NECESSARY
IF MAILED
IN THE
UNITED STATES

Fold carefully along this line.

Interactive Mathematics Program
Year 4
Comment Form

Please take a moment to provide us with feedback about IMP. If you have comments or suggestions about Year 4, we'd like to read them. Once you've filled out this form, all you have to do is fold it and drop it in the mail. We'll pay the postage. Thank you!

Your Name _____

School _____

School Address _____

City/State/Zip _____

Phone _____

1. List any comments about the IMP *Year 4* student text.

   _____
   _____
   _____
   _____
   _____

2. List any comments about the teacher's guide for _____.

   _____
   _____
   _____
   _____

3. Do you have any other comments about IMP *Year 4* or any suggestions for improving the student text or teacher material?

   _____
   _____
   _____
   _____
   _____

Thank you for taking the time to fill out this comment form.
Please return completed forms to:
   Editorial—IMP, Key Curriculum Press, Box 2304, Berkeley, CA 94702.
   You can fold this form as shown on the back and it becomes a postage-paid self-mailer!

Fold carefully along this line.

BUSINESS REPLY MAIL
FIRST CLASS MAIL   PERMIT NO. 151   BERKELEY, CA

POSTAGE WILL BE PAID BY ADDRESSEE

NO POSTAGE
NECESSARY
IF MAILED
IN THE
UNITED STATES

KEY CURRICULUM PRESS
Innovators in Mathematics Education

P.O. Box 2304
Berkeley, CA 94702-9983
Attn: Editorial—IMP

Fold carefully along this line.